I0787957

Semiconductor Electrolyte Interface and Photoelectrochemistry

Editor:

R. Mantz
Army Research Office
Durham, North Carolina, USA

Sponsoring Divisions:

 Energy Technology

 Physical and Analytical Electrochemistry

Published by
The Electrochemical Society
65 South Main Street, Building D
Pennington, NJ 08534-2839, USA
tel 609 737 1902
fax 609 737 2743
www.electrochem.org

ECStransactions ™

Vol. 25, No. 42

Copyright 2010 by The Electrochemical Society.
All rights reserved.

This book has been registered with Copyright Clearance Center.
For further information, please contact the Copyright Clearance Center,
Salem, Massachusetts.

Published by:

The Electrochemical Society
65 South Main Street
Pennington, New Jersey 08534-2839, USA

Telephone 609.737.1902
Fax 609.737.2743
e-mail: ecs@electrochem.org
Web: www.electrochem.org

ISSN 1938-6737 (online)
ISSN 1938-5862 (print)
ISSN 2151-2051 (cd-rom)

ISBN 978-1-56677-817-6 (PDF)
ISBN 978-1-60768-167-0 (Softcover)

Printed in the United States of America.

Preface

The papers included in this issue of *ECS Transactions* were originally presented in the symposium "Semiconductor Electrolyte Interface and Photoelectrochemistry", held during the 216th meeting of The Electrochemical Society, in Vienna, Austria from October 4 to 9, 2009.

***ECS Transactions*, Volume 25, Issue 42**
Semiconductor Electrolyte Interface and Photoelectrochemistry

Table of Contents

Preface   *iii*

Electron Transport Processes in Dye-Sensitized Solar Cells Made of One-Dimensional   1
Titania Nanoscale Materials
     M. Adachi and Y. Mori

Photo-anodic Polarization Behavior of TiO_2 with Dye Sensitization for Electrochemical   9
Polymerization of Pyrrole
     J. Kawakita

Interfacial Electrochemistry of ZnO: An Influence of the Surface Polarity?   21
     A. Joudrier, F. Levy, N. Simon, M. Bouttemy, F. Levy, G. Feuillet and A. Etcheberry

Junction Effect on the Photocatalytic Activity of Mixed-Phase TiO_2 Nanoparticles   29
     A. Di Paola, M. Bellardita, L. Palmisano and F. Parrino

Dye-Sensitized Solar Cells Using Localized Surface Plasmon of Gold and Silver   37
Nanoparticles with Comb-shaped Block Copolymer
     M. Enomoto, K. Taniguchi and M. Ihara

Photoelectrochemistry of Hematite Thin Films   49
     H. Wang and J. A. Turner

Photo-Induced Alcohol Electro-Reforming for H_2 Production   63
     A. K. Seferlis and S. G. Neophytides

Photoelectrolysis of Water in Tj-a-Si Solar Cell Biased CM-n-TiO_2 $|$ $|$ Pt and in Monolithic   73
Self-Driven n-TiO_2 - Mn_2O_3 Coated Tj-a-Si $|$ $|$ Pt Photoelectrochemical Cells
     M. Frites, W. Ingler Jr. and S. U. Khan

Low Reflectance Surface Observed on InP Porous Structures after Photoelectrochemical   83
Etching
     T. Sato, N. Yoshizawa, H. Okazaki and T. Hashizume

Author Index   89

Facts about ECS

The Electrochemical Society (ECS) is an international, nonprofit, scientific, educational organization founded for the advancement of the theory and practice of electrochemistry, electrothermics, electronics, and allied subjects. The Society was founded in Philadelphia in 1902 and incorporated in 1930. There are currently over 7,000 scientists and engineers from more than 70 countries who hold individual membership; the Society is also supported by more than 100 corporations through Corporate Memberships.

The technical activities of the Society are carried on by Divisions. Sections of the Society have been organized in a number of cities and regions. Major international meetings of the Society are held in the spring and fall of each year. At these meetings, the Divisions and Groups hold general sessions and sponsor symposia on specialized subjects.

The Society has an active publications program that includes the following.

Journal of The Electrochemical Society — JES is the peer-reviewed leader in the field of electrochemical and solid-state science and technology. Articles are posted online as soon as they become available for publication. This archival journal is also available in a paper edition, published monthly following electronic publication.

Electrochemical and Solid-State Letters — ESL is the first and only rapid-publication electronic journal covering the same technical areas as JES. Articles are posted online as soon as they become available for publication. This peer-reviewed, archival journal is also available in a paper edition, published monthly following electronic publication. It is a joint publication of ECS and the IEEE Electron Devices Society.

Interface — *Interface* is ECS's quarterly news magazine. It provides a forum for the lively exchange of ideas and news among members of ECS and the international scientific community at large. Published online (with free access to all) and in paper, issues highlight special features on the state of electrochemical and solid-state science and technology. The paper edition is automatically sent to all ECS members.

Meeting Abstracts (formerly Extended Abstracts) — Abstracts of the technical papers presented at the spring and fall meetings of the Society are published on CD-ROM.

ECS Transactions — This online database provides access to full-text articles presented at ECS and ECS-sponsored meetings. Content is available through individual articles, or as collections of articles representing entire symposia.

Monograph Volumes — The Society sponsors the publication of hardbound monograph volumes, which provide authoritative accounts of specific topics in electrochemistry, solid-state science, and related disciplines.

For more information on these and other Society activities, visit the ECS website:

www.electrochem.org

Electron Transport Processes in Dye-Sensitized Solar Cells Made of One-Dimensional Titania Nanoscale Materials

Motonari Adachi and Yasushige Mori

Doshisha University, Faculty of Science and Engineering,
1-3 Miyakodani Tatara, Kyotanabe 610-0321, Japan

It is elucidated that electron diffusion rate in titania electrode in dye-sensitized solar cells should be very rapid, i.e., highly crystallized TiO_2 one-dimensional materials are needed. Three kinds of TiO_2 nanoscale materials (1,2) were synthesized, applied for dye-sensitized solar cells, and all cells made of these three materials showed high light-to-electricity conversion efficiency, around 9%. Electrochemical impedance spectroscopy is very useful to investigate electron transport processes in dye-sensitized solar cells (5). Strong distribution of electron density induced by distribution in irradiation light under steady state conditions must be taken into consideration to determine parameters participating electron transport processes accurately and to evaluate the quality of the cells correctly.

Necessity of highly crystallized one-dimensional TiO_2 nanoscale materials for fabricating highly efficient dye-sensitized solar cells

Highly crystallized one-dimensional titania nanoscale materials are the most promising materials for an electrode of dye-sensitized solar cells (DSSCs). We succeeded in the preparation of titania nanorods (TR) (1) (see Figure 1), network structure of titania nanowires (2) and one-dimensional titania nanochains (see Figure 2), which have been newly synthesized. We applied these materials for DSSCs.

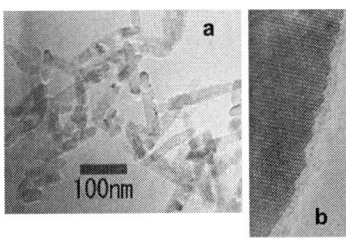

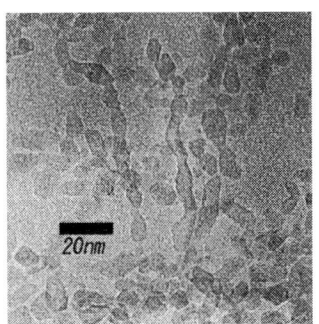

Figure 1. TEM images of high crystalline titania nanorods (TR) with adsorbed dyes. (a) Low magnified image and (b) High resolution image near the edge of TR.

Figure 2. TEM image of titania nanochains.

First, let us consider the reason why highly crystallized one-dimensional titania materials are needed. Figure 3a shows a typical Nyquist plot obtained by electrochemical impedance spectroscopy (EIS). Total direct current (dc) resistance is given by the length

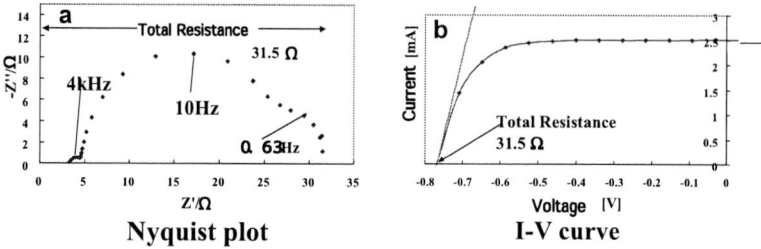

Figure 3. (a) Typical Nyquist plot obtained by EIS, (b) I-V curve for the same cell.

from 0 to the point at ω=0 on the real axis as shown by Figure 3a. This fact is confirmed later by reproduction of I-V curve using measured total dc resistances at various bias voltages as shown in Figure 4. Total dc resistance is also obtained from the slope of the tangent line at the point of Voc. (Figure 3b) When the total dc resistance becomes small, the slope becomes steep, and the fill factor becomes larger, resulting in a high light-to-electricity conversion efficiency. Thus, the total dc resistance should be small. However, the largest arc of around 10 Hz in Figure 3a represents the resistance of recombination reactions between electrons in the titania electrode and I_3^- ions in the electrolyte. Small total dc resistance means small resistance for recombination reactions, indicating rapid reaction rate of recombination. Thus, small total dc resistance seems an obstacle for attainment of highly efficient solar cells. But, whether electrons in the titania electrode are properly collected by the transparent conducting glass electrode or react with I_3^- ions in the electrolyte by recombination reactions is determined by the ratio of the resistance for the transport rate to the conducting glass electrode against the resistance for the recombination reactions. When the resistance for the transport rate to the conducting glass electrode is much smaller than that of the recombination reactions, almost all electrons are properly collected by the conducting glass electrode. This means that the transport rate of electrons in the titania electrode should be very rapid, indicating that we need nice titania materials with high electron transport rate, i.e., highly crystallized one-dimensional nanoscale TiO_2 materials are needed.

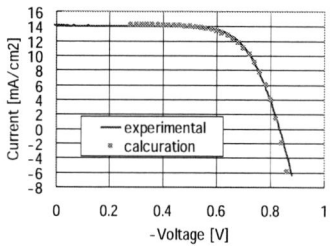

Figure 4. Reproduction of I-V curve by total dc resistances at various bias voltages.

Solid line in Figure 4 shows experimentally obtained I-V curve under illumination. The square keys show calculated curve based on the observed total dc resistances at various bias voltages and the following relationship between current density and voltage,

$$di = \frac{dV}{R_t} \tag{1},$$

where R_t stands for total dc resistance. The calculated curve reproduces experimentally obtained I-V curve very well, confirming that the total dc resistances can be determined accurately from Nyquist plot of EIS analysis.

Next, we present highly crystallized one-dimensional titania nanoscale materials are effective to attain high light-to-electricity conversion yield. As shown in our previous paper (1), network structure of single crystal-like titania nanowires can be synthesized successfully by the oriented attachment mechanism. We attained 9.33 % conversion efficiency with complex titania electrode made of titania nanowires and P-25. Recently, we attained the same conversion efficiency 9.33 % using different electrolyte, i.e., 0.6M 1-butyl 3-methyl imidazolium iodide, 0.1M guanidium thiocyanate, 0.05M I_2, 0.5M tert-butylpyridine in a mixture of acetonitrile and valeronitrile (85:15) for a complex titania electrode made of titania nanowires, titania nanoparticles (3-5 nm in diameter, see reference 3,4) and P-25. (Figure 5)

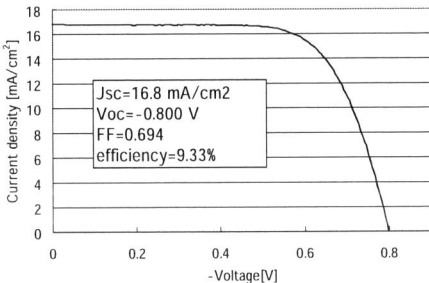

Figure 5. I-V curve obtained for a cell with a complex electrode composed of network structure of single-crystal-like titania nanowires, titania nanoparticles and P-25.

In our previous paper (1), we used an electrolyte composed of 0.1 M of LiI, 0.6 M of 1,2-dimethyl-3-n-propylimidazolium iodide, 0.05 M of I_2, 1 M of 4-tert-butylpyridine in methoxyacetonitrile and got 9.33 % conversion efficiency with short circuit current density Jsc=19.2 mA/cm^2, open circuit voltage Voc=0.72 V and fill factor 0.675. In the recent results, Voc value 0.8 V is larger than that of previous one 0.72 V, because guanidium thiocyanate decreased redox potential of I^-/I_3^- in the electrolyte. Unfortunately, we got lower short circuit current density Jsc=16.8 mA/cm^2 than that of our previous one, and the same efficiency was obtained.

Highly crystallized titania nanorods (TR) have been synthesized by hydrothermal process using blockcopolymer (F127) and surfactant cetyltrimethylammonium bromide (CTAB) as a mixed template. (2) As shown in Figure 1a, TR with 100-300 nm in length and 20-30 nm in diameter was obtained. A high-resolution TEM (HRTEM) image of single TR (Figure 1b) shows that titanium atoms align perfectly in titania anatase crystalline structure with no lattice defect, and the surface of TR is facetted with the TiO_2 anatase {101} faces. The fringes are {101} planes of anatase TiO_2 with a lattice spacing of about 0.351 nm, which agrees with the value recorded in JCPDS card. The highly crystallized titania nanorods prepared successfully were used to fabricate a titania electrode of DSSCs. The complex electrodes were made by the repetitive coating-calcining process: 3 layers of titania nanoparticles (3-5 nm in diameter, see reference 3,4) were first coated on FTO conducting glass, followed by 8 layers of mixed gel composed of titania nanorods and titania nanoparticles. A high light-to-electricity conversion efficiency of 8.93 % was achieved.

We have newly synthesized titania nanochains as shown in Figure 2. The formation procedure was the same as that of titania nanorods (see reference 2), except no use of ethylendiamine and adjusting pH of the aqueous solution of tetra-isopropylorthotitanate (TIPT), F127 and CTAB to acidic conditions; pH= 1.3 - 5.0. Highly crystallized titania nanoparticles with diameter of around 10 nm combine with each other and make chains. The obtained white solid product was mixed with spherical titania nanoparticles (3-5 nm in diameter) synthesized using F127 reported in our previous paper (3, 4) to fabricate titania film electrodes. The I-V curve of the cell is shown in Figure 6. The obtained light-to-electricity conversion yield of the cell was 9.2%.

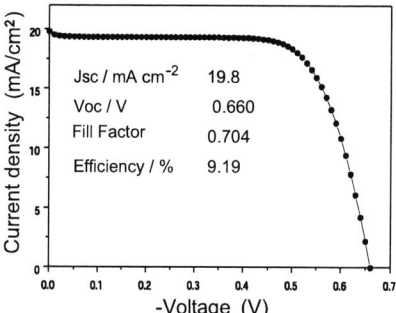

Figure 6. I-V curve obtained for the cell composed of one-dimensional chains of titania nanoparticles mixed with fine titania nanoparticles (3 - 5 nm in diameter).

All three kinds of one-dimensional titania nanoscale materials show high light-to-electricity conversion yield around 9%, suggesting strongly that highly crystallized one-dimensional titania materials are essentially important for attainment of high efficient dye-sensitized solar cells.

Effect of strong distribution of electron density on EIS analysis

It is important to analyze electron transport processes in DSSCs to attain further improvement in conversion efficiency. Electron impedance spectroscopy (EIS) is very useful method for measurement and analysis of electron transport processes in DSSCs. We found that the reliable values of parameters relating to electron transport in dye-sensitized solar cells can be determined from measured spectra by EIS when careful analysis of measured spectra was done based on the classification and clarification of the impedance equation obtained from the solution of differential equations which were derived based on the behavior of carriers in the solar cell. (5) The requisites for making highly efficient dye-sensitized solar cells were proposed, and especially the high diffusion rate of electrons in TiO2 electrode was pointed out as the key factor for achievement of high efficiency.

Recently, we find that there are some essential problems concerning with EIS analysis. So far impedance analysis in dye-sensitized solar cells have been carried out in many papers based on the assumption of constant photoelectron generation rate of electrons regardless of the distance from the interface between TiO_2 and transparent conducting glass to the interface between TiO_2 and the electrolyte under open circuit condition. (6, 7) Kern et al. derived impedance equation based on the same assumption of constant photoelectron generation rate $\overline{G}$ of electrons as given by Eq.(2) under open circuit condition. (8)

$$\overline{G} = \frac{1}{L} \int_0^L \alpha \Phi_0 \exp(-\alpha x) dx \qquad (2)$$

However, incident light should be absorbed efficiently in the solar cell, and the intensity of light should decrease exponentially to a very low intensity level. So, there must be strong distribution of light in TiO_2 electrode, resulting in strong distribution of photoelectron generation.

Let derive the equation which take the distribution of photoelectron generation into consideration. The assumption $k_1 \gg k_2$ by Kern et al. results in $N_0 \gg n_0$, where k_1 and k_2 represent trapping and detrapping rate constant of electrons, and N_0 and n_0 stand for the equilibrium electron density of trap and conduction band, respectively. Thus, almost all electrons are in the trap state. Electrons in the trap state detrap to the conduction band and diffuse with the diffusion coefficient D_{cb} for the period proportional to k_2/k_1, where D_{cb} represents diffusion coefficient of electrons in the conduction band. Thus, electrons in the trap state are regarded as diffusing charges with the diffusion constant $D_{eff}=D_{cb}(k_2/k_1)$. Electrons in the trap state are assumed to react with I_3^- with second order reaction rate according to Kern et al. k_{eff} represents rate constant for recombination from trap state. Differential equation including distribution of photoelectron generation under steady state condition was derived as Eq. (3).

$$D_{eff} \frac{\partial^2 N_0}{\partial x^2} - k_{eff} N_0{}^2 + \alpha \Phi_0 \exp(-\alpha x) = 0 \qquad (3)$$

where, α stands for the effective absorption coefficient. Calculated steady state electron distribution under open circuit condition is given in Figure 7a. The electron density distribution is not constant uniform value, but varies with distance x. Strong distribution is induced by the strong distribution of light intensity. Contribution of each term in Eq.(3), i.e., diffusion, recombination and generation terms are shown in Figure 7b.

However, under non-steady state conditions, the differential equation does not include photo-generation term, because incident light irradiates steadily, including no non-steady component. The differential equation is derived as Eq.(4).

$$\frac{\partial \tilde{n}}{\partial t} = D_{eff} \frac{\partial^2 \tilde{n}}{\partial x^2} - k_{eff}\tilde{n}, \quad \tilde{n} = \Delta n(x)\exp(i\omega t) \tag{4}$$

where, $\tilde{n}$ represent non-steady oscillating electron density and Δn stands for amplitude of oscillation. Eq.(4) holds for both dark and illumination conditions. Boundary conditions are also same in both cases. Thus, the same impedance equation is obtained for under dark and under illumination coOnditions. Are impedance data obtained under dark same

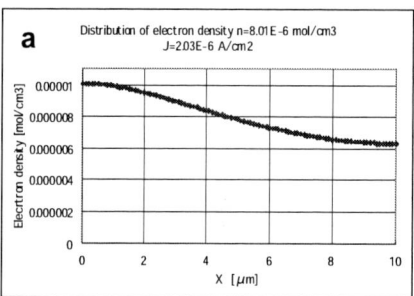

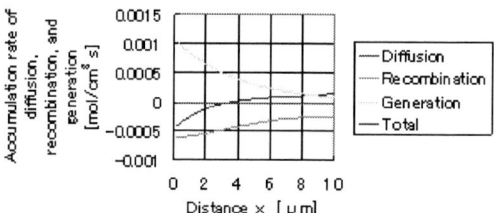

Figure 7. (a) Steady state electron density distribution under Voc condition.
(b) Contribution of each term in Eq.(3); diffusion, recombination and generation.
Calculation conditions: $D_{eff} = 1 \cdot 10^{-5}$ cm^2/s, $\alpha = 2500$ s^{-1}, $k_{eff} = 6 \cdot 10^6$ cm^{3}mol^{-1}s^{-1}.

as those obtained under illumination? Steady electron distribution is quite different. We carried out EIS experiments to get the answer for this question. Figure 8 shows Nyquist plots under dark and illumination conditions. The obtained spectrum under illumination is quite different from that obtained under dark. Wang et al. (9) obtained similar experimental results. Their explanation is that I_3^- ions are produced near the interface of TiO_2 electrode under illumination, resulting in decrease in resistance for recombination reactions. We further carried out experiments with variation of bias voltages. When the voltage becomes 0.5 V, the resistance for dark condition is about 100 times larger than that under illumination condition. Two order difference is too large for the explanation due to the effect of I_3^- concentration difference. Figure 10 shows relationship between

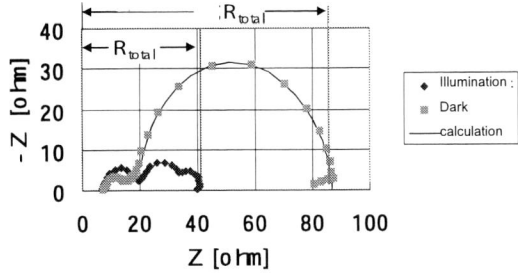

Figure 8. Nyquist plots under dark and illumination conditions.

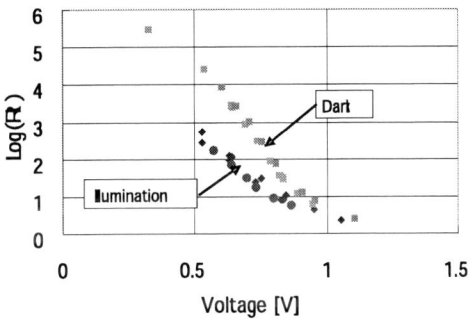

Figure 9. Relationship between resistance for recombination reactions and bias voltages.

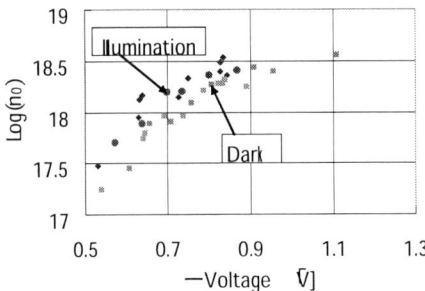

Figure 10. Relationship between electron density and bias voltages.

electron density and bias voltages. Electron densities under illuminatio conditions are 40-50% larger than those of dark conditions. This differences in electron density would induce big change in resistance of recombination reactions, especially in the case of low electron density such as the case of 0.5 V. We also found that the larger electron density under illumination conditions result in the larger electron diffusion coefficients than those under dark conditions.

Materials and Methods

Fuluorine doped tin oxide (FTO) transparent conducting glass was purchased from NSG Co., Ltd. F127 [(PEO)$_{106}$-(PPO)$_{70}$-(PEO)$_{106}$] was obtained from BASF. TIPT, CTAB, and ethylenediamine were purchased from Tokyo Kasei Organic Chemicals.

The synthetic methods of titania nanoscale materials used in this work were shown in our previous papers (1-4) as mentioned in the text.

Summary

1. All three kinds of one-dimensional titania nanoscale materials show high light-to-electricity conversion yield around 9%. This suggests strongly that highly crystallized one-dimensional titania materials are essentially important for attainment of high efficient dye-sensitized solar cells.
2. Under steady state conditions, strong distribution of electron density is induced by strong distribution of irradiating light even under open circuit condition. Impedance spectrum under dark is quite different from that of under illumination because of the difference in steady state electron density. It is very important to take into consideration of electron density distribution in the titania electrode and to establish the improved procedure to determine parameters participating electron transport process accurately and simply and also to evaluate the quality of the whole cell.

References

1. M. Adachi, Y. Murata, J. Takao, J. Jiu, M. Sakamoto, and F. Wang, *Journal of The American Chemical Society*, **126**, 14943 (2004)
2. J. Jiu, S. Isoda, F. Wang, and M. Adachi, *Journal of Physical Chemistry B*, **110**, 2087 (2006)
3. J. Jiu, F. Wang, M. Sakamoto, J. Takao and M. Adachi, *Journal of The Electrochemical Society*, **151**, A1653 (2004)
4. J. Jiu, S. Iszoda, M. Adachi, and F. Wang, *Journal of Photochemistry and Photobiology A: Chemistry*, **189**, 314 (2007)
5. M. Adachi, M. Sakamoto, J. Jiu, Y. Ogata, and S. Isoda, *Journal of Physical Chemistry B*, **110**, 13872 (2006)
6. J. van de Lagemaat, N. G. Park, A. J. Frank, *J. Phys. Chem. B*, **104**, 2044 (2000)
7. L. Peter, *J. Phys. Chem. C*, **111**, 6601-6612 (2007)
8. R. Kern, R. Sastrawan, J. Ferber, R. Stangl and J. Luther, *Electrochim. Acta*, **47**, 4213 (2000)
9. Q. Wang, J-E. Moser, M. Grätzel, *J. Phys. Chem. B* **109**, 14945 (2005)

Photo-anodic Polarization Behavior of TiO₂ with Dye Sensitization for Electrochemical Polymerization of Pyrrole

Jin Kawakita[a, b]

[a] Advanced Photovoltaics Center, National Institute for Materials Science (NIMS), Tsukuba 305-0047, JAPAN
[b] World Premier International Research Center Initiative on Materials Nanoarchitectonics (MANA), Tsukuba 305-0047, JAPAN

Solidification of the electrolyte in dye-sensitized solar cell is important to improve the safety and the reliability by the low risk in leakage of the electrolyte from the cell. The hole transport material is a promising candidate for the solid electrolyte. The optimized interfacial structure between the photo-anode and the hole transport material is not still unclear. Formation of organic polymer through photo-anodic polymerization on the photo-anode is attractive with respect to filling of the polymer to fine porous structures in the photo-anode and to controlling the electronic structure of the polymer as the p-type semiconductor. In this paper, formation behavior of polypyrrole was studied through the photo-electrochemical measurements for the aggregates of anatase TiO₂ with dye sensitization and characterization of the reaction products to get the guideline for designing the organic polymer. The results showed that polypyrrole was formed localizedly on the TiO₂ layer and that polymerization and side reaction took place simultaneously.

Introduction

One of the important issues for practical use of the dye-sensitized solar cell (DSSC) is to ensure the long-term reliability and stability (1). Although liquid electrolyte in the state of the art shows comparatively high conductivity, it may evaporate and the cell performance will fall when it is used for a long term. Moreover, it is a problem that the liquid electrolyte will leak by the damage of the cell etc. Substitution of the liquid electrolyte with a hole transport material such as inorganic and organic solid compound, ionic gel, and organic semiconductor is an essential solution to fabricate solid DSSC because of no overpotential for charge transfer and mass transfer such as $3I^-/I_3^-$ couple in the liquid electrolyte (2-5).

Aggregate of nano TiO_2 particulates covered with dye takes advantage of the larger surface area for photo-electric conversion on DSSC. The oxidative polymerization is a good way to fill the pores among nano TiO_2 particulates with p-type organic semiconductor as the hole transport material. In that way, anodic current necessary for the oxidative polymerization of organic monomer is obtained through hole generated by the photo-excitation phenomenon on the dye-sensitized TiO_2. This polymerization, what is called, photo-anodic polymerization was studied on formation of polypyrrole on TiO_2 in aqueous solution with respect to photo-electrochemistry (6). In addition, the photo-anodic polymerization takes advantage of controlling the electronic structures of p-type organic semiconductor obtained. Some attempts with polypyrrole (PPy) and poly(3.4-ethylenedioxythiophene) (PEDOT) prepared by photo-anodic polymerization were reported as solid electrolyte in DSSC and the conversion efficiency was considerably smaller than that observed with the usual liquid electrolytes (7-10). It is necessary to clarify both the photo-electric conversion and the charge transfer processes for the improvement of the solid DSSC with p-type organic semiconductors. Prior to that, it is important to research the photo-electrochemical process to control structures and properties of the organic semiconductors.

In this paper, PPy was selected as the organic semiconductor and photo-anodic polarization behavior of TiO_2 with dye sensitization was studied in the non-aqueous solution containing pyrrole (Py) as the precursor for polymerization.

Experimental

The electrochemical measurements were carried out by using the three electrode cell with an open window with 10 mm in diameter to set the working electrode, which was prepared as follows. Slurry containing TiO_2 (anatase, 20 nm in size, Ishihara Sangyo Kaisha Ltd.) was coated on a conductive transparent oxide (TCO) glass substrate and was heated at 723 K for 30 min. After the heat treatment, the specimen was immersed at ambient temperature for 24 hours in ethanol solution containing the dye known as N719; cis-bis(isothiocyanato)bis(2,20-bipyridyl-4,40-dicarboxylato)-ruthenium(II)bis-tetrabutylammonium (Solaronix SA). The electrolyte was solution of Pyrrole (Py) as a precursor for polymerization and tetraethylammonium p-toluenesulfonate (TEAPTS, Kanto Chemical Co., Inc.) as a dopant and supporting salt in ethanol and polyethyleneglycol (PEG, Mw 200, Kanto Chemical Co., Inc.). The counter electrode was a platinum plate and the reference electrode was Ag/AgCl/saturated KCl solution. In this paper, the electrode potential is expressed against SHE. The visible light was introduced with a solar simulator (SX-UI 501XQ; Ushio Inc.) from the backside of the working electrode.

Linear sweep voltammetry was carried out from the immersion potential to 2.0 V at a sweep rate of 25 mV·min^{-1}. Chrono-potentiometry was carried out at a constant anodic current of 0.75~75 µA·cm^{-2}. All the electrochemical measurements were carried out with equipment by Hokuto Denko Corp. (Model HZ-5000).

The specimens were observed with scanning electron microscopes (SEM: JEOL, Model JSM-5400 and JSM-6500) and analyzed with energy dispersive X-ray spectrometer (EDX). The FT-IR spectra of the specimens were obtained with a spectrophotometer (Spectrum GX-R; Perkin Elmer) in multiple reflection mode.

Results and discussion

Fig. 1 shows anodic polarization curves of TiO$_2$ with dye adsorption in ethanol with and without pyrrole under illumination. Additional curve in dark was obtained regarding the TCO glass electrode. When the TCO glass plate was used as the working electrode in the electrolyte containing Py and TEAPTS solution in the dark, small current was observed around 10^{-6} A·cm^{-2} from the initial potential to 1 V, and then a rapid increase in current was observed. This rapid increase in current was due to formation of PPy on the TCO surface through anodic polymerization, resulted from formation of PPy film on the TCO glass electrode after anodic polymerization. Under illumination, the immersion potentials was less noble than that observed in the dark because of occurrence of photo-electromotive force derived from generation of photo-excited electron in the conduction band of TiO$_2$. In addition, the electrode potential was scanned from the initial potential to the anodic direction while current was rapidly increased. With Py in the electrolyte, the immersion potential was further less negative and the current was higher at a certain potential up to -0.1 V than that without Py. This can be explained by large fraction of the anodic current consumed for formation of PPy through a photo-electrochemical reaction. The current level between -0.1 and 1.0 V under illumination was almost constant approximately one hundred times higher than that in the dark and showed a similar behavior regardless from the presence of Py in the electrolyte. This steady current against the scanned potential is characteristic of limitation in reaction rate by movement of hole generated after photo-excitation. A similar current behavior can be explained by preferential contribution of the electrolyte to the photo-anodic reaction such as decomposition rather than formation of PPy. The detailed comparison showed that the current with pyrrole in the electrolyte was smaller than that without Py at a certain potential between -0.1 and 1.0 V. This can be explained by smaller surface area of the working electrode effective for the reaction with the electrolyte, resulted from the electrode covered with PPy formed, as shown below. When Py was contained in the electrolyte, current was increased above 1.0 V. This can be explained by superimposing of current for electrochemical polymerization of Py on the TCO glass substrate and/or on PPy connected with the substrate through the TiO$_2$ layer. The current, however, was smaller than the value observed on the bare TCO glass substrate. This is explained by decrease in diffusion of the electrolyte in the fine pores composed of stacking TiO$_2$ particles.

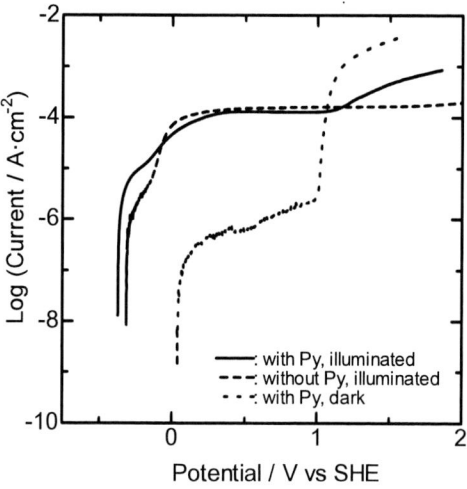

Figure 1. Anodic polarization curves of TiO$_2$ with dye adsorption in ethanol with and without pyrrole under illumination. Additional curve in dark was obtained regarding the TCO glass electrode.

Fig. 2 shows anodic polarization curves of the specimens in PEG solution under some conditions. Additional curve in dark was obtained regarding the TCO glass electrode. Similarly to the polarization behavior in the ethanol solution, the slight current below 10^{-6} A·cm^{-2} was observed on the TCO glass electrode under the dark condition in the electrolyte containing Py and TEAPTS in PEG. Above 1.0 V, the current was rapidly increased. This is due to formation of PPy through anodic polymerization. Under illumination, the immersion potential was shifted to the less negative direction also in case of PEG as a solvent because of photo electromotive force mentioned above. PEG showed less negative immersion potential under illumination than ethanol. This was due to smaller cathodic current in PEG than ethanol. PEG showed the following behavior similar to ethanol, 1) the less negative immersion potential with Py in the electrolyte, 2) the rapid increase in current from the initial potential by scanning of the potential to the anodic direction and 3) the higher current with Py than without Py in the electrolyte at a certain potential up to -0.25 V. These facts implied formation of PPy by photo-anodic polymerization. The current with Py in PEG was smaller than that in ethanol, indicating smaller polymerization rate in the PEG solution. The current level between -0.25 and 1.6 V under illumination was around one hundred times higher than that in the dark and showed a similar behavior regardless from the presence of Py. The current was gradually increased, which was different from the result in ethanol, i.e. almost constant. Above 1.6 V, the current was increased when Py was contained in the electrolyte. This was due to superimposing of the current for electrochemical polymerization of Py on the TCO substrate. A gradual increase in current between -0.25 and 1.6 V and shifting in breaking potential of electrochemical polymerization on the TCO glass substrate to the positive direction from 1.0 V to 1.6 V suggested small reaction rate of polymerization in the PEG solution, in other words, large overpotential. This is because PEG has a larger viscosity ca. 4 mm²·s⁻¹ than ethanol 1 mm²·s⁻¹, leading to the higher resistance in movement of the electrolyte.

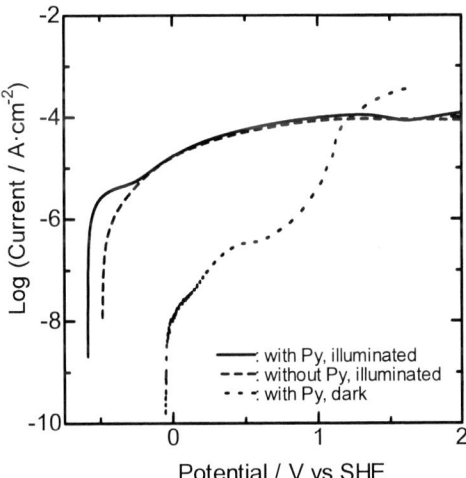

Figure 2. Polarization curves of TiO_2 with dye adsorption in polyethylene glycol with and without pyrrole under illumination. Additional curve in dark was obtained regarding the TCO glass electrode.

Fig. 3 shows changes in potential with the duration time upon photo-anodic polarization at a constant current in ethanol solution. At 7.5 $\mu A \cdot cm^{-2}$, the potential level seems similar regardless from the presence of Py. This is because the energy level of hole generated by photo-excitation is low enough to be able to precede polymerization of PPy and decomposition of the solvent simultaneously although the electrode potential indicates the energy level of the electron in TiO_2. At 75 $\mu A \cdot cm^{-2}$, overpotential becomes smaller in the presence of Py than in the absence of Py. This is because the polymerization reaction has smaller polarization resistance than side reaction such as decomposition of the solvent. The potential went up gradually with time and rapidly after some duration time, indicating that surface of TiO_2 was covered with PPy and that subsequent photo-excitation reaction on TiO_2 is not possible. In addition, the constant potential was observed around 0.9 V. This is due to formation of PPy on the TCO glass substrate through the electrochemical polymerization.

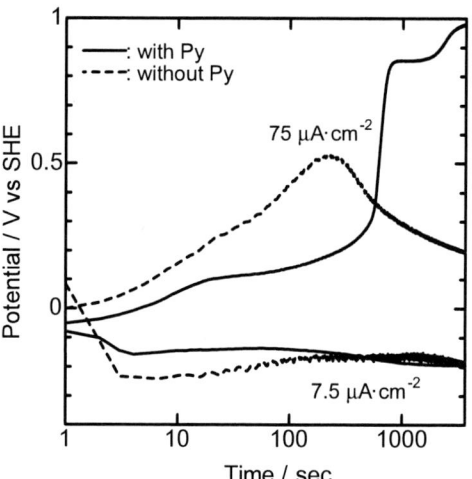

Figure 3. Changes in potential with duration time upon photo-anodic polarization at constant current in ethanol solution.

Fig. 4 shows changes in potential with the duration time upon photo-anodic polarization at a constant current in the PEG solution. At 7.5 $\mu A \cdot cm^{-2}$, an initial potential step and subsequent potential increase with the different slope suggested possible formation of PPy through the photo-anodic reaction. In the second potential step, the similar potential level was obtained, indicating preferential contribution of the electrolyte to the photo-anodic reaction such as decomposition of the solvent. At 75 $\mu A \cdot cm^{-2}$, overpotential in PEG became larger than that in ethanol, implying high resistance for the anodic reaction. This is explained by the difference in viscosity between ethanol and PEG, as mentioned above.

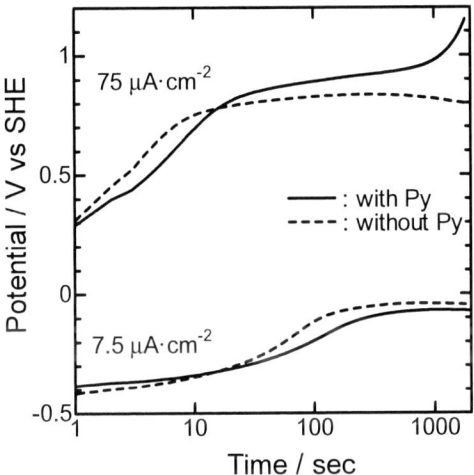

Figure 4. Changes in potential with duration time upon photo-anodic polarization at constant current in polyethylenglycol solution.

Fig. 5 shows SEM images of surface appearance on the specimens after galvanostatic polarization with different charges in ethanol solution containing Py under illumination. A table beside each image shows the EDX result of the corresponding specimen. At a charge of 1.35 $mC \cdot cm^{-2}$ (= 0.75 $\mu A \cdot cm^{-2} \times 30$ min), something like splat was observed localizedly, as seen in Fig. 5a. At a charge of 27.0 $mC \cdot cm^{-2}$, its coverage area was increased, as seen in the image of Fig. 5b, and the EDX result showed the increase in atomic ratio of carbon element. It is considered that this splat was PPy formed by photo-anodic polymerization and that the rest of surface area in TiO_2 might contribute to react with the electrolyte. The localized formation of PPy can be explained by the deduction of higher activation energy for nucleation of PPy on the TiO_2 surface or of inhomogeneous arrival of light through the TiO_2 layer to the interface between TiO_2 and the electrolyte. In larger charge of 135 mC·cm^{-2}, almost all the surface of TiO_2 was covered with the splat, as seen in Fig. 5c.

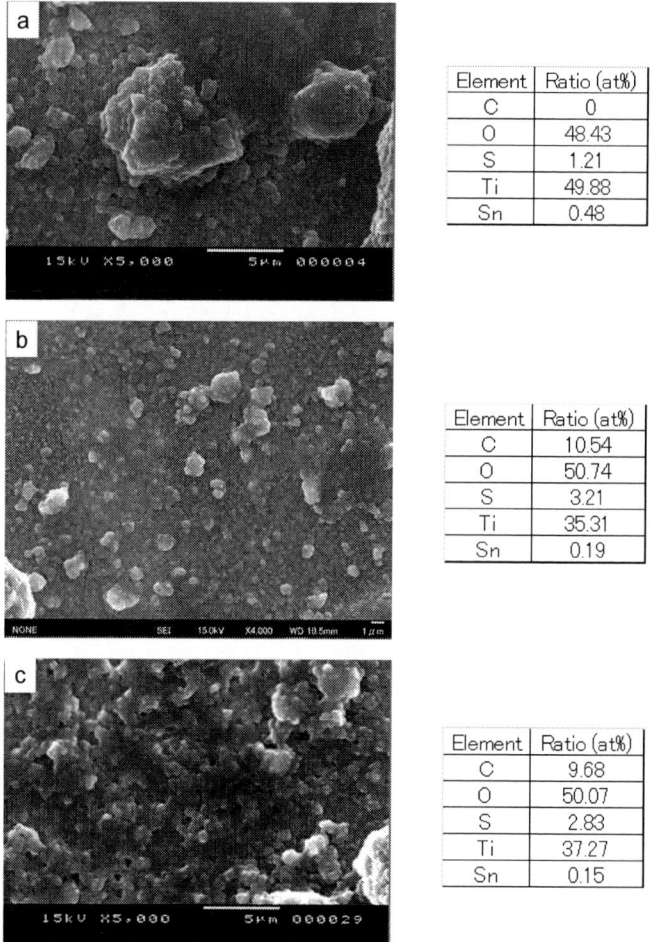

Element	Ratio (at%)
C	0
O	48.43
S	1.21
Ti	49.88
Sn	0.48

Element	Ratio (at%)
C	10.54
O	50.74
S	3.21
Ti	35.31
Sn	0.19

Element	Ratio (at%)
C	9.68
O	50.07
S	2.83
Ti	37.27
Sn	0.15

Figure 5. SEM images of surface appearance on TiO_2 with dye adsorption after galvanostatic polarization with different charges at (a) 0.75 µA·cm^{-2} × 30 min, (b) 7.5 µA·cm^{-2} × 60 min and (c) 75 µA·cm^{-2} × 30 min in ethanol solution containing Py under illumination. Table beside each image shows EDX result of corresponding specimen.

Fig. 6 shows SEM images of surface appearance in the specimens after galvanostatic polarization with different charges in PEG solution containing Py under illumination. A table beside each image shows the EDX result of the corresponding specimen. Similarly to the case in ethanol, splat was observed on the surface of TiO_2 localizedly and increased with the charge, as seen in Figs. 6a and 6b. The coverage area by splat, in other words, the amount of splat formed, however, was considerably smaller at the same order in charge than that prepared in ethanol (see Figs 5a and 5b). The carbon element was not detected through EDX at charges of 0.75 $\mu A \cdot cm^{-2} \times$ 30 min and 7.5 $\mu A \cdot cm^{-2} \times$ 60 min. On the surface of the specimen after reaching the higher potential above 1.0 V, agglomeration of spherical particles were observed, as shown in Fig. 6c, and the EDX result showed the considerable increase in atomic ratio of carbon element. These can be explained by electrochemical polymerization of pyrrole on resulting polypyrrole connected with the substrate through the TiO_2 layer.

Element	Ratio (at%)
C	0
O	56.14
S	0.23
Ti	43.31
Sn	0.13

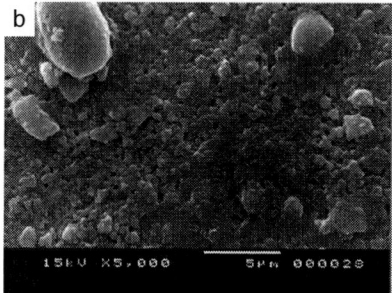

Element	Ratio (at%)
C	0
O	65.49
S	0.02
Ti	31.54
Sn	2.94

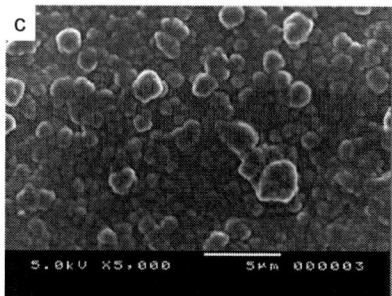

Element	Ratio (at%)
C	22.71
O	58.29
S	0.69
Ti	10.02
Sn	13.23

Figure 6. SEM images of surface appearance on TiO_2 with dye adsorption after galvanostatic polarization with different charges at (a) 0.75 $\mu A \cdot cm^{-2} \times$ 60 min, (b) 7.5 $\mu A \cdot cm^{-2} \times$ 60 min and (c) 75 $\mu A \cdot cm^{-2} \times$ 60 min in polyethyleneglycol solution containing Py under illumination. Table beside each image shows EDX result of corresponding specimen.

Fig. 7 shows IR spectra of the specimens. Polarized specimens are compared to TiO_2 on TCO substrate, and subsequently dye-adsorbed TiO_2. When polarized at a constant current of 7.5 $\mu A \cdot cm^{-2}$ for 60 min in ethanol solution under illumination, additional peaks appeared obviously around 685, 1010, 1035 and 1120 cm^{-1} in wave number accompanying with the peaks attributed to the TCO glass substrate and TiO$_2$ and dye. These peaks are consistent with those observed on PPy prepared on the TCO glass substrate through electrochemical polymerization at a constant current of 75 μA·cm^{-2} for 60 min in the dark. When polarized in PEG solution under the same condition as in ethanol solution, the peaks attributed to PPy cited above could not be observed regardless from the current density and time. Furthermore, longer the duration time is, smaller the peaks attributed to the TCO substrate and TiO$_2$ became. From this result, other organic compounds might be formed, taking into account formation of the splat in PEG solution. For instance, those were polypyrrole with different chemical structure in amorphous state or with low molecular weight. Further characterization of chemical structure is in progress.

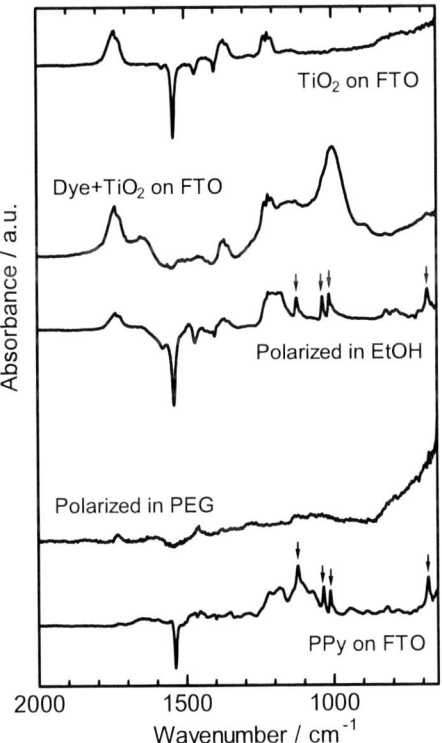

Figure 7. IR spectra of specimens
Arrows show peaks assigned to polypyrrole.

Conclusion

We revealed that formation behavior of polypyrrole through photo-anodic polarization on TiO_2 with dye sensitization in the solution of pyrrole/tetraethylammonium p-toluenesulfonate/ ethanol or polyethyleneglycol.

The reaction potential was divided into the three regions; 1) preferential occurrence of photo-anodic polymerization of polypyrrole, 2) simultaneous occurrence of photo-anodic polymerization and side reaction such as decomposition of the solvent, and 3) anodic polymerization of polypyrrole.

Polyppyrole through photo-anodic reaction was formed localizedly on TiO_2 with dye adsorption. The rate in photo-anodic reaction with the solvent is higher than photo-anodic polymerization rate of polypyrrole. The photo-anodic current is considered to depend on resistance of the electrolyte related to the viscosity.

Acknowledgments

Mr. A. Nozaki and Ms. C. Iso are greatly appreciated for their experimental assistance.

References

1. B. O'Regan and M. Grätzel, *Nat.*, **353**, 737 (1991).
2. B. O'Regan and D.T. Schwartz, *Chem. Mater.*, **10**, 1501 (1998).
3. U. Bach, D. Lupo, P. Comte, J. E. Moster, F. Weissortel, J. Salbeck, H. Spreitzer, M. Gratzel, *Nat.*, **395**, 583 (1998).
4. S. Murai, S. Mikoshiba, H. Sumino, T. Kato and S. Hayase, *Chem. Comm.*, 1534 (2003).
5. Y. Saito, T. Kitamura, Y. Wada and S. Yanagida, *Synth. Met.*, 131 (2002) 185.
6. M. Okano, K. Itoh, A. Fujishima and K. Honda, *J. Electrochem. Soc.*, **134** (1987) 837.
7. K. Murakoshi, R. Kogure, Y. Wada, S. Yanagida, K. Murakoshi, *Chem. Lett.*, **26**, 471 (1997).
8. T. Kitamura, M. Maitani, M. Matsuda, Y. Wada and S. Yanagida, *Chem. Lett.*, (2001) 1054.
9. R. Cervini, Y. Cheng and G. Simon, *J. Phys. D*, **37**, 13 (2004).
10. Y. Saito, N. Fukuri, R. Senadeera, T. Kitamura, Y. Wada and S. Yanagida, *Electrochem. Comm.*, **6**, 71 (2004).

Interfacial electrochemistry of ZnO: an influence of the surface polarity?

A.-L. Joudrier[a], N. Simon[a], L. Santinacci[a], M. Bouttemy[a], F. Levy[b], G. Feuillet[b] and A. Etcheberry[a]

[a]Institut Lavoisier de Versailles UMR-CNRS 8180 Université de Versailles 45 Ave des Etats Unis 78035 Versailles, France
[b]DOPT/SIONA CEA-LETI CEA-LETI, 17 rue des Martyrs 38054 Grenoble, France

The question of the nature of the polar surface on ZnO single-crystal is studied using different approaches. XPS, etching rate determinations and electrochemical measurements are performed to collect information about the nature of the polar character of these faces. Different behaviors are detected that can be related to polarity effects.

Introduction

Zinc oxide (ZnO) is among II-VI compounds one of the more important (1) with its wide bandgap (3.37 eV at 300 K) and its exciton energy of 60 meV at room temperature. It presents a lot of very interesting physical and chemical properties that can be extended toward applications. This importance is reflected by the great interest of possible applications in optoelectronic devices as lasers, light emitting diodes, optical coatings, and solar cells. A strength of this material is that large and high quality single-crystal are now commercially available that allows to grow homo-epitaxial layers, then ZnO-based devices. Its weakness is attaining reliable and robust p-type doping. This goal which is still controversial needs studies to obtain good p-type growth. So to perform good, homo-epitaxial growths it needs to develop efficient characterizations of ZnO substrates then epitaxial layers. A very interesting tool is the electrochemical approach. Interfacial electrochemistry can provide a lot of information about the surface chemistry and the relevant behaviors. Electrochemistry on ZnO is a long scientific story. It supports pioneer works on the topic (2) and it is obvious that this material has given a large contribution to the fundamentals of the semiconductor electrochemistry (3) as flat band potential pH dependence (4) current doubling effect (5) charge transfer process (6). However, even if a lot of ZnO electrochemical behaviors are established, it stays a material for electrochemical studies even it was strangely forgotten during last decades as mono-crystalline electrode. The electrochemical characterization, of single-crystal with oriented face, which can be used as substrate for epitaxy, has been the purpose of this work. Zinc oxide presents a wurtzite structure, each Zn atom is surrounded in a tetrahedric disposition by four O atoms. Along the c-axis Zn and O planes alternate and the surface termination of the crystal can be either Zn-face (ZnO(0001)) or O-face (ZnO(000-1)). In fact, the surface configurations of the ZnO single crystal are very complex (7) even if the notion of polar surface is well established on this material through very specific behaviors. However polarity of (0001) faces is clearly evidenced by specific experimental behaviors which obviously differ depending on the nature of the exposed polar face. GaN epilayer growth is of better quality on O-face (8). Variation of the quality of the layer is also detected for ZnO homo-epitaxy. The more obvious evidence of face polarity influence was observed on the difference of chemical instability and reactivity in acidic or basic solutions. This reactivity features has been pointed out at the beginning of the electrochemical studies on ZnO (4, 9) and used as a proof of the actual polarity of the (0001) faces. This difference of reactivity is detected by very different evolutions of the surface morphology as dissolution progresses. This method is now currently used for epitaxial or technological purposes (10). Nevertheless this difference is partially understood even dissolution mechanism differences have been suggested (9). Etching differences don't mean directly polar properties but only morphology variations happening when significant amount of matter are released in solution.

The challenge in our present study is to work on the notion of polar surface by developing an experimental approach that links XPS and electrochemical considerations. As very high quality of the substrates are now available, it allows a quantitative approach of an eventual influence of the very different surface chemistries related to the polar surfaces, onto XPS, chemical and the electrochemical behaviors. Before immersion the polarity of the initial faces has been clearly established by XPS and ARXPS that give specific responses which depend on the face polarity (11).

Experimental

Oriented single-crystals are provided by CRYSTEC. During electrode preparation the studied polar face is protected touching a clean glass plane during a room temperature in rear-soldering. For XPS analyses samples are introduced as received, after chemical or electrochemical treatments. After dipping of surfaces in solution with or without electrochemical treatments, released Zn in solution is determined by F&GF-AAS (Thermo Electron) dosages. XPS measurements are performed on a Thermo Electron VG-Escalab 220i-XL and ARXPS measurements are performed on a Thermo Electron THETA probe. Electrochemical experiments are performed using a classical three electrode configuration with a PARSAT 2273 potentiostat.

Results and discussion

Characterization of as received polar faces

The as-received polar faces present by AFM slight but clear morphological differences as shown in figure 1. The same samples are also analyzed using high resolution XPS for which clear differences appear for Zn_{2p}, O_{1s} and Zn_{LMM} contributions (fig. 2).

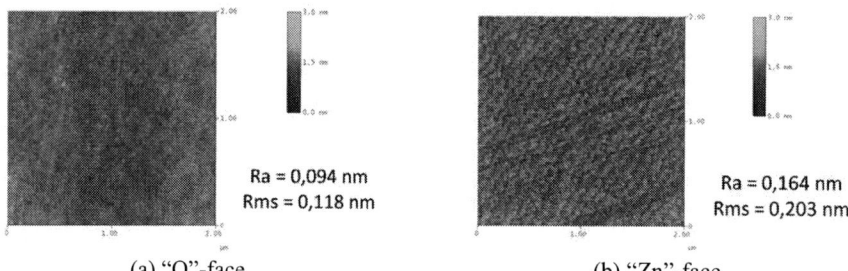

Ra = 0,094 nm
Rms = 0,118 nm

Ra = 0,164 nm
Rms = 0,203 nm

(a) "O"-face   (b) "Zn"-face
Figure 1: AFM images of as-received O-face (a) and Zn-face (b) on ZnO single-crystals.

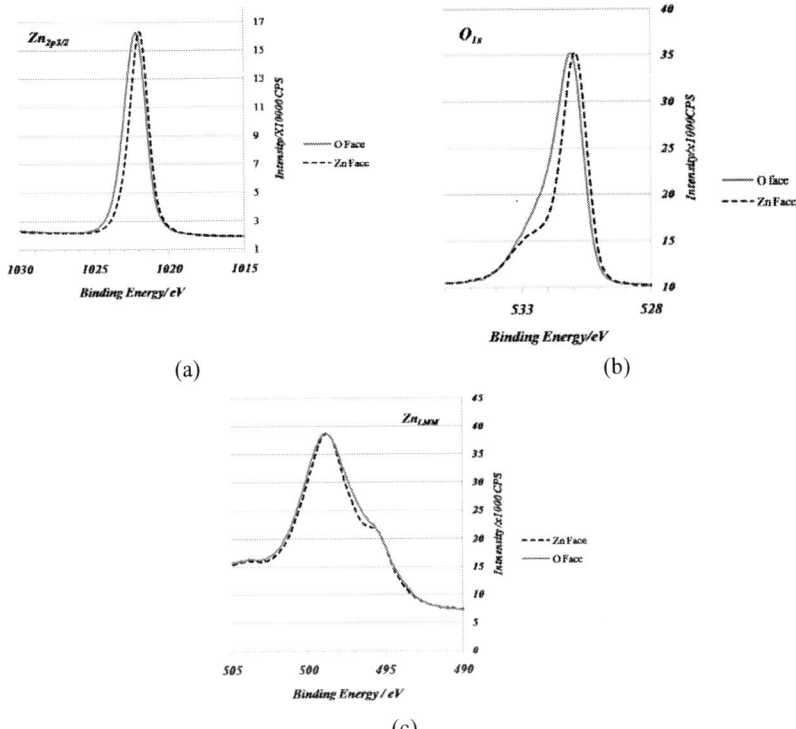

(a)   (b)

(c)

Figure 2: Comparison for Zn-face and O-face of Zn_{2p} (a), O_{1s} (b) and Zn_{LMM} (c) spectra.

These initial characterizations show that the preparation of polar faces is able to discriminate between Zn and O polarities. If the AFM detects only a morphological difference, XPS demonstrates that clear differences can be assumed independently of their physical or chemical origin. O_{1s} presents significant variation of the energy distribution but more interesting Zn signals also. To complete these observations it must be considered that the analyses of adventitious carbon contamination are very close and very reproducible allowing a direct exploitation of the O and Zn spectra differences. So polar effects can be distinguished and the agreement of any following characterization with the trends presented in fig. 1 and fig. 2 can be used as a test of preservation of the initial polar character after chemical, ageing, thermal or electrochemical treatments.

Chemical reactivity of the polar faces

As already mentioned the O and Zn faces present very different behaviors as soon as they undergo a chemical etching. Figure 3 shows the evolution of the surfaces morphology after dipping in HCl 1M. Note that similar features are detected in H_3PO_4 or H_2SO_4. So it is clear that from a very similar initial state, the surfaces evolve towards very different states under chemical etching as already observed by others authors (12).

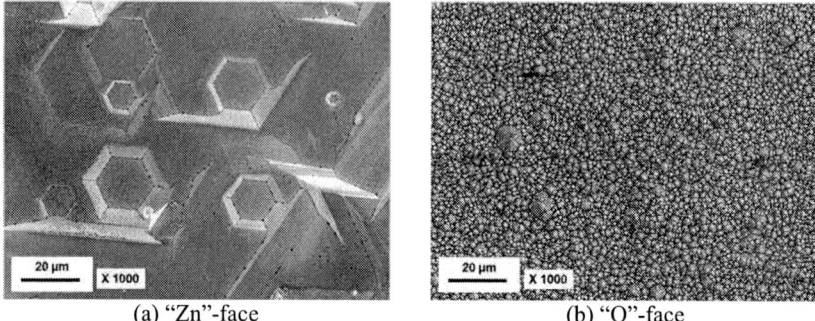

(a) "Zn"-face   (b) "O"-face

Figure 3: Evolution of the surface morphologies after dipping in HCl 1M during several hours (20 h) - (a) Zn-face, (b) O-face - the initial situations were as shown in figure 1.

The more interesting thing about the chemical reactivity is detected when the reactivity is followed by real time measurements of the dissolution. Dosages of Zn amount released in solution are performed for different immersion times and pH conditions (fig. 4).

HCl (10^{-3} M)

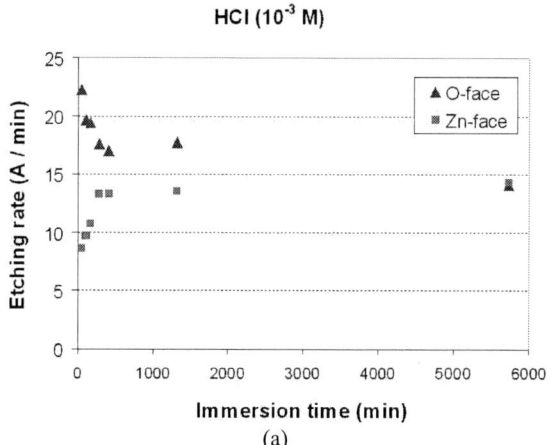

(a)

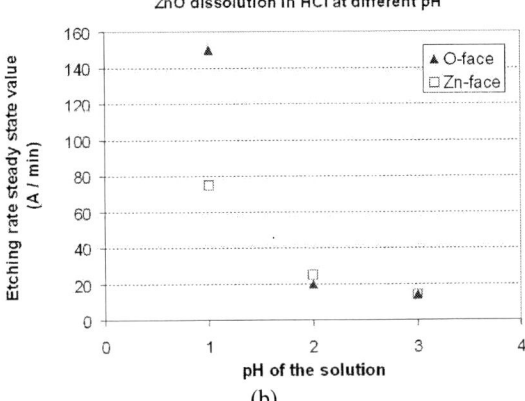

(b)

Figure 4: Study of the dissolution of the ZnO (0001) and (000-1) faces in HCl: (a) influence of the face polarity on the instantaneous etching rate, (b) influence of the pH on the steady state etching rate. The etching rates are determined by AAS dosages of the Zn.

In slight acidic solutions the more interesting fact is that the instantaneous etching rate depends, at the beginning of the immersion, on the nature of the polar face. This very important fact is demonstrated by sequential Zn dosage (fig. 4a). After sufficient immersion time the etching rate tends toward the same limit as if the polarity of the face was lost. The influence of the pH shows also that this same limit is not reached for more acidic solutions (fig. 4b). In this case the etching rates between O and Zn faces stay different. The rate is always higher for O-face. So the detection of an instantaneous etching rate differentiates as for XPS, information that can be linked directly to a polarity effect. The last point associated to the chemical reactivity in acidic medium concerns the XPS response of the surfaces after etching. The initial XPS differentiation is generally lost with a final state that is intermediate. It is perfectly evident for low pH conditions, it is more difficult to conclude for intermediate pH. For example ultra pure water rinsing doesn't displace the initial response.

So this work on the instability of ZnO in aqueous media demonstrates that immersion in solution must be taken with care due to potential perturbation as soon as immersion is done. The history of the sample must be known to provide a pertinent interpretation of the results.

Electrochemical behavior of (0001) polar faces

When the electrochemical experiments are performed on surfaces prepared by a preliminary etching in acidic solution the responses on Zn or O faces are quasi-identical. The I(V) curves present very large without current domains (several volts) limited in the negative bias range by a classical hydrogen evolution.

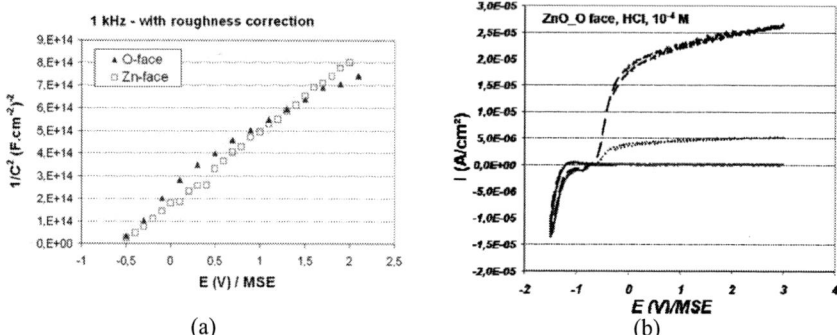

(a)   (b)

Figure 5: (a) Mott-Schottky plot performed on O and Zn faces after pre-etching or long immersion. (b) I(V) in the dark or under illumination performed on O or Zn faces after pre-etching or long immersion.

Under illumination the $I_{ph}(V)$ curves depends on the incident light intensity with a characteristic potential distribution that are also the same for O and Zn faces (fig. 5b). The photo dissolution that is associated to this anodic current has been kinetically studied by Zn dosage in solution. A $2e^-$ mechanism has been established on Zn and O faces (5).

For the same conditions of surface preparation the $C^{-2}(V)$ present a perfect linear Mott-Schottky plots with reproducible flat band potential that doesn't depend on the nature of the face (fig. 5a). So ZnO single-crystals present electrochemical experiments as soon as the reactivity has been induced that are very similar whatever the nature of the initial face.

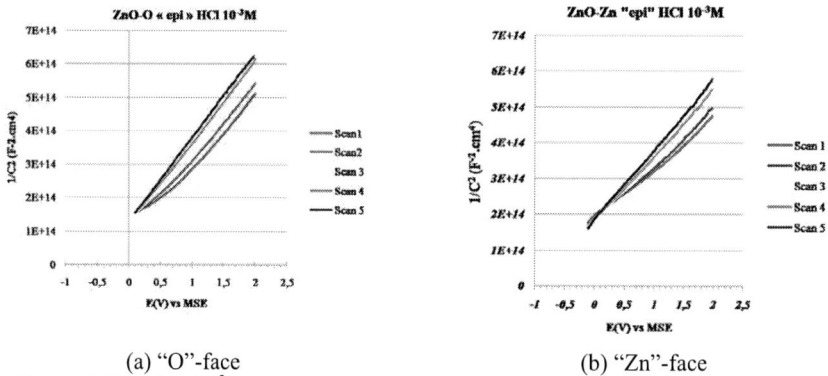

(a) "O"-face   (b) "Zn"-face

Figure 6: Transient $C^{-2}(V)$ curves on "fresh" O (a) or Zn (b) faces performed in low etching solutions (HCl 10^{-3}M)

The situation changes when the initial ZnO surface is kept as close as possible of its initial state. To respect these conditions, it needs that the preparation of the initial surface doesn't include etching step even very weak. Only rising with ultra pure water and methanol can be performed keeping XPS responses close as the ones presented in figure

2. In this case, typical evolutions of the $C^{-2}(V)$ curves with progressive evolutions are observed as soon as immersion is performed in a very diluted HCl solution for which very low etching rate have been established.

As shown in figure 6, similar shifts of the $C^{-2}(V)$ curves are detected. Critical times for these C(V) measurements are shorter than the ones detected for etching rate evolutions on polar faces. So the observations given here for C(V) evolutions must be considered for surfaces with clear different etching rate. The polarity of the initial surface can be considered as maintained. The global slopes are very close as expected for samples cut on the same crystal then in agreement with a same doping density. The interesting and original features concern the extrapolations of the $C^{-2}(V)$ which differ clearly depending on the nature of the polar face. Even it is difficult to give accurate values it is obvious that flat band potential on Zn faces are more negative than the one obtained on O faces.

Conclusion

The question of the detection of the polar character of 0001 and 000-1 faces of ZnO has been studied on high quality single-crystals. It is obvious that XPS data present differences on as received polar surfaces that can be interpreted as consequence of the polar nature of the prepared surfaces. For low etching conditions, it is also pointed out that polar surfaces undergo different initial etching processes, which again can be interpreted as a consequence of the polarity of the faces. The last original point is associated to the C(V) experiment that are directly related to the energy diagrams of the respective interfaces. For initial surfaces without perturbation sufficient experimental trends are observed that suggest an effect of the polarity on the energy diagrams of the interfaces obtained from polar surface of ZnO single-crystals.

Acknowledgments

This work has been supported by Institut Carnot program

References

1. *ZnO bulk, thin films and Nanostructures,* Ed. C. Jagadish and S.J. Pearton, Elsevier Oxford UK, (2006)
2. J.F. Dewald, *Bell System Tech. J.*, **39**, 615, (1960)
3. S.R. Morrison in *Electrochemistry at semiconductor and oxidized metal electrode* Plenum Press, New York, (1980)
4. F. Lohmann, *Ber. Bunsenges. Phys. Chem.*, **70**, 428, (1966)
5. S.R. Morrison, T. Freund, *J. Chem.Phys.*, **47**, 1543, (1967)
6. P.A. Kohl, A. Bard, *J. Am. Chem. Soc.*, **99**, 7531, (1977)
7. C. Wöll, *Progress in Surface Science*, **82**, 55, (2007)
8. D.C. Look, *Mater. Sci. Eng.*, **B80**, 383, (2001)
9. B. Pettinger, H.-R. Schöppel, T. Yokoyama, H. Gerischer, *Ber. Bunsenges. Phys. Chem.*, **78**, 1024, (1974)
10. F. Hamdani, M. Yeadon, J. Smith David, H. Tang, W. Kim, A.E. Botchkarev, J.M. Gibson, A.Y. Polyakov, M. Skowronski, H. Morkoc, *J. Appl. Phys.*, **83**, 983 (1998)
11. L. Zhang, D. Wett, R. Szargan, T. Chassé, *Surf. Interface Anal.*, **36**, 1479, (2004)
12. A. N. Mariano, R. E. Hanneman, *J. Appl. Phys.*, **34** (2), 384, (1963)

28

Junction Effect on the Photocatalytic Activity of Mixed-Phase TiO$_2$ Nanoparticles

A. Di Paola, M. Bellardita, L. Palmisano and F. Parrino

"Schiavello-Grillone" Photocatalysis Group, Dipartimento di Ingegneria Chimica dei Processi e dei Materiali, Università di Palermo, Viale delle Scienze, 90128 Palermo, Italy.

Active TiO$_2$ photocatalysts were prepared under mild experimental conditions by thermohydrolysis of TiCl$_4$ in pure water at 100 °C. The preparation method is very simple and does not require the use of expensive thermal or hydrothermal treatments. Depending on the TiCl$_4$/H$_2$O ratio, pure rutile, binary mixtures of anatase and rutile or anatase and brookite, or ternary mixtures of anatase, brookite and rutile, can be obtained. 4-nitrophenol photodegradation was used to evaluate the photoactivity of the various powders. The high photocatalytic activity of the mixed samples was explained by the presence of junctions among different polymorphic TiO$_2$ phases that allows an improved charge separation of the photogenerated electron-hole pairs, hindering their recombination.

Introduction

TiO$_2$ is the most studied photocatalyst for environmental applications because of its high efficiency, nontoxicity, chemical and biological stability, and low cost (1). TiO$_2$ exists in three different crystalline habits: anatase, brookite and rutile. Anatase is generally accepted to be a photocatalyst more efficient than rutile and brookite but mixtures of the TiO$_2$ polimorphic phases often show photocatalytic activities superior than those of the pure phases (2). The coupling of different semiconductors possessing different energy levels for their corresponding conduction and valence bands is usually beneficial to enhance the photocatalytic activity (2, 3). The very active commercial Degussa P25, that is frequently used as a benchmark in heterogeneous photocatalysis, consists of a mixture of anatase and rutile.

The present paper reports on the preparation and characterization of highly active TiO$_2$ photocatalysts prepared by thermohydrolysis of TiCl$_4$ in pure water at 100 °C. The reactivity of the various samples was tested for the photocatalytic degradation of 4-nitrophenol and compared to that of Degussa P25. The flat band potential and the band gap values of anatase, brookite and rutile were also determined. The results are discussed by taking account of the relative positions of the energy bands of the three TiO$_2$ phases.

Experimental

Preparation and characterization of the samples

1 ml of TiCl$_4$ (Fluka 98%) was slowly added to different volumes of distilled water at room temperature. The solutions obtained after continuous stirring were heated in closed bottles and aged at 100 °C in an oven for 48 h. The bottles were allowed to cool and the resultant solids were recovered using a vacuum pump at 55 °C. The crystal phase

composition of the catalysts was determined by X-ray diffraction (XRD) measurements using a modified Rietveld method (4). The specific surface areas (SSA) were obtained by nitrogen physisorption experiments.

Brookite was separated by peptization from a binary mixture of brookite and rutile, obtained by thermolysis of a solution of $TiCl_4$ in diluted HCl ($TiCl_4:H_2O:HCl$ volume ratio 1:300:20) at 100 °C for 48 h (2). After repeated washings with water, a dispersion of brookite particles formed while the rutile phase remained as precipitate. The sol containing the brookite particles and the precipitate of rutile were separately dried under vacuum at 55 °C. Anatase was prepared by boiling an aqueous solution of $TiCl_4$ ($TiCl_4:H_2O$ volume ratio 1:50) for 2 h. After removal of the supernatant liquid, the solid was dried under vacuum at 55 °C.

The flat-band potentials of anatase, brookite and rutile were determined by measuring the photovoltage as a function of the suspension pH in the presence of methyl viologen dichloride (5). The band gap values of the three phases were obtained by diffuse reflectance spectra measurements.

Photoreactivity experiments

A Pyrex batch photoreactor of cylindrical shape containing 0.5 L of aqueous suspension was used. A 125 W medium pressure Hg lamp (Helios Italquartz, Italy) was immersed within the photoreactor and the photon flux emitted by the lamp was $\Phi_i = 13.5$ $mW \cdot cm^{-2}$. O_2 was continuously bubbled for ca. 0.5 h before switching on the lamp and throughout the occurrence of the photoreactivity experiments. The amount of catalyst was $0.6 \ g \cdot L^{-1}$ and the initial 4-nitrophenol (BDH) concentration was $20 \ mg \cdot L^{-1}$. Samples of 5 mL were withdrawn at fixed intervals of time with a syringe, and the catalyst was separated from the solution by filtration. Some drops of a 1 M NaOH solution were added before filtration to cause agglomeration of the particles. The quantitative determination of 4-nitrophenol was performed by measuring its absorption at 315 nm with a spectrophotometer Shimadzu UV-2401 PC.

Results and discussion

Figure 1 shows the X-ray diffraction patterns of the solids formed by thermohydrolysis of $TiCl_4$ in water at 100 °C. The composition of the powders depended on the $TiCl_4/H_2O$ ratio and binary or ternary mixtures of the three polimorphs were prevalently produced. The percentages of anatase, brookite and rutile present in the various samples after 48 h of heating are reported in Table I. The average particle sizes of all the phases present in the various samples were in the range 2-10 nm.

TABLE I. Influence of the $TiCl_4/H_2O$ ratio on the relative content (wt%) of anatase (A), brookite (B) and rutile (R), specific surface area (SSA) and initial reaction rate (r_0) of the samples.

sample	[TiCl$_4$] [mol·L^{-1}]	wt %			S.S.A [m^2·g^{-1}]	$r_0 \cdot 10^9$ [mol·L^{-1}·s^{-1}]
		A	B	R		
TiO$_2$(1:5)	1.48			100	87.5	31
TiO$_2$(1:10)	0.81	19		81	108.0	63
TiO$_2$(1:25)	0.34	53		47	201.5	61
TiO$_2$(1:35)	0.25	70	19	11	196.1	47
TiO$_2$(1:50)	0.17	66	21	13	196.6	58
TiO$_2$(1:75)	0.12	65	28	7	216.1	76
TiO$_2$(1:100)	0.09	66	28	6	219.1	67
TiO$_2$(1:150)	0.06	77	23		208.3	33

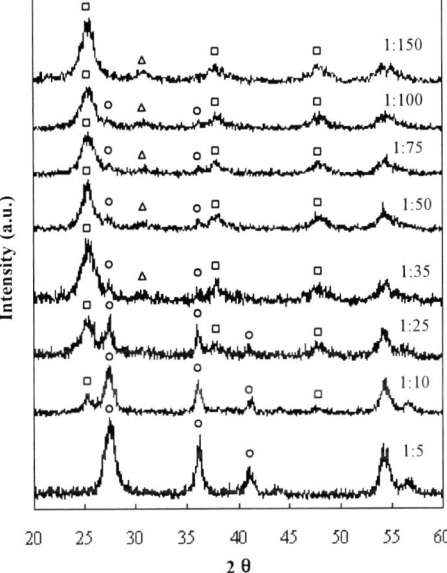

Figure 1. XRD patterns of the solids obtained by thermohydrolysis of TiCl$_4$ in water at 100 °C for 48 h. (□) anatase; (o) rutile; (△) brookite.

The H_2O content is the factor which determines the nature of the crystalline phases and their relative proportions. The fraction of rutile decreases with dilution and this is consistent with the results of previous works indicating that low pH values favour the formation of rutile (6, 7). Only rutile was obtained when the $TiCl_4/H_2O$ ratio was 1:5. The $TiCl_4/H_2O$ ratio plays an important role in the competition between the formation of anatase and rutile, but has little effect on brookite formation.

Figure 2 shows a microscopic procedure proposed to explain the formation of the different TiO$_2$ polymorphic species. According to Zheng *et al.* a low concentration TiCl$_4$ solution contains a large amount of disperse $[Ti(OH)_2(OH_2)_4]^{2+}$ octahedral complexes (8). As a consequence of hydrothermic treatments, the octahedra link together by olation, through dehydration reactions between aquo and hydroxo ligands. Rutile type nuclei are developed if the $[Ti(OH)_2(OH_2)_4]^{2+}$ monomers combine by sharing equatorial edges, whereas anatase or brookite type nuclei form if the monomers combine by sharing apical edges. Further growth proceeds by formation of linear chains from the rutile type nuclei or of skewed chains from the anatase or brookite type nuclei. All the three polymorphic phases can be obtained contemporaneously and the composition of the mixtures depends on the $TiCl_4/H_2O$ ratio and by the reaction time.

When the concentration of $TiCl_4$ is high, the solution should prevalently contain $[TiO(OH_2)_5]^{2+}$ momomers that can combine by olation only by sharing equatorial edges. In this case, rutile crystallites are developed (8).

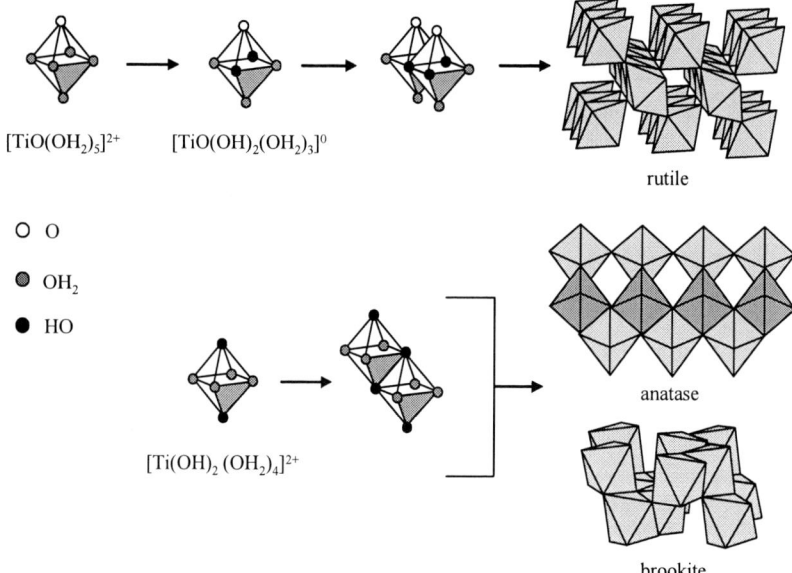

Figure 2. Possible reaction pathways for the formation of rutile, anatase and brookite starting from octahedral complexes: $[TiO(OH)_2(OH_2)_3]^0$ and $[Ti(OH)_2(OH_2)_4]^{2+}$(8).

The photocatalytic activity of the samples, as well their composition, depends on the $TiCl_4/H_2O$ ratio. Table I reports the values of the initial degradation rate of 4-nitrophenol, r_0, calculated from the initial slope of the concentration versus time profiles. The mixed systems revealed an enhanced photoactivity compared with that of the pure TiO_2 polymorphic phases and some samples were more active than Degussa P25. The most efficient samples consisted of a ternary mixture of anatase, brookite and rutile.

The high photocatalytic activity of the binary or ternary mixtures can be explained by the presence of junctions among different polymorphic TiO_2 phases possessing different energy levels for their corresponding conduction and valence bands (3). A current hypothesis of the enhanced activity of mixed phases is the vectorial transfer of electrons from a semiconductor to another, leading to more efficient electron-hole separation and greater catalytic reactivity (9).

Diffuse reflectance measurements allowed to determine the band gap of pure anatase, brookite and rutile. The estimated values were 3.05 eV for anatase, 3.26 eV for brookite and 2.98 eV for rutile, respectively. The band gaps of the mixtures of two or three phases were lower than those of anatase and brookite and near to that of rutile.

The values of the flat-band potentials of the three pure phases were determined by the slurry method proposed by Roy *et al.* (5), measuring the variation of the photovoltage with the pH of suspensions of the powders in the presence of an electron acceptor. Figure 3 shows the effect of pH on the photovoltage developed on irradiation of anatase, brookite or rutile suspensions. From the value of the inflection point (pH_0), the flat band potential at pH 7 was calculated by the equation:

$$E_{FB} (pH = 7) = E_{MV}^{2+/+\cdot} + 0.059 (pH_0 - 7) \qquad [1]$$

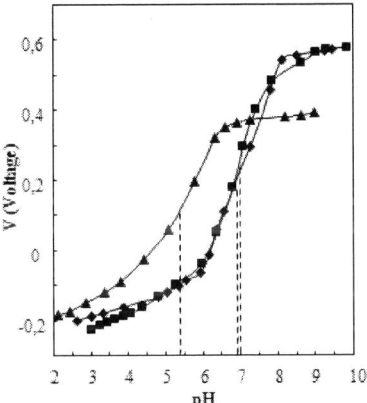

Figure 3. Effect of pH on the photovoltage developed on irradiation of (♦) anatase, (■) brookite and (▲) rutile suspensions in the presence of methyl viologen dichloride.

The values obtained were - 0.45 V, - 0.46 V and - 0.37 V for anatase, brookite and rutile, respectively.

Assuming that the difference between flat band potential and conduction band edge is negligible, it is possible to locate the valence band edge of the three semiconductors by adding the band gap energy to the flat-band potential value. Figure 4 shows the relative positions of the energy bands of anatase, brookite and rutile, at pH = 7. Anatase and brookite differ in the position of their valence band since brookite has a slightly larger band gap energy.

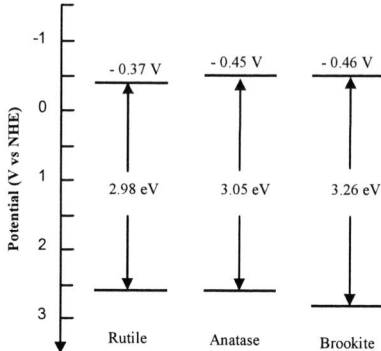

Figure 4. Electrochemical potentials (versus NHE) of the band edges of anatase, brookite, and rutile at pH = 7.

From the relative position of the band edges, it can be expected that electrons pass from the more cathodic conduction band of one semiconductor to the more anodic conduction band of another component. Another explanation recently proposed by Hurum *et al.* (9) that cannot be excluded is that electron transfer occurs from the more anodic conduction band of one phase (e.g. rutile) to trapping sites of another polymorph (e.g. anatase) which are lower in energy than the conduction band.

The photoactivity of the mixed powders depends on the composition, the aggregates size and the distribution of the various junctions among different phases. The small size of the particles and the intimate contact with each other are crucial factors to increase the efficiency of the samples.

Conclusion

Highly active photocatalytic TiO_2 samples can be synthesized by thermohydrolysis of $TiCl_4$ in water at 100 °C. The most efficient samples consisted of a ternary mixture of anatase, brookite and rutile. The presence of junctions among different polymorphic TiO_2 phases enhances the separation of the photogenerated electron-hole pairs, reducing their recombination.

The composition of the samples can be easily tailored by simply varying the $TiCl_4/H_2O$ volume ratio. A proper $TiCl_4/H_2O$ volume ratio is important for the synthesis of a ternary mixture of anatase, brookite and rutile which exhibits the highest photoactivity.

Acknowledgments

This work was financially supported by the Ministero dell'Istruzione, dell'Università e della Ricerca (Roma).

References

1. A. Fujishima, K. Hashimoto and T. Watanabe, *TiO₂ Photocatalysis: Fundamentals and Applications*, Bkc, Tokyo (1999).
2. A. Di Paola, G. Cufalo, M. Addamo, M. Bellardita, R. Campostrini, R. Ceccato, M. Ischia and L. Palmisano, *Colloid. Surf. A: Physicochem. Eng. Aspects*, **317**, 366 (2008).
3. N. Serpone, P. Maruthamuthu, P. Pichat, E. Pelizzetti and H. Hidaka, *J. Photochem. Photobiol. A: Chem.*, **85**, 247 (1995).
4. L. Lutterotti, R. Ceccato, R. Dal Maschio and E. Pagani, *Mater. Sci. Forum,* **278-281**, 87 (1998).
5. A. M. Roy, G. C. De, N. Sasmal and S. S. Bhattacharyya, *Int. J. Hydrogen Energy*, **20**, 627 (1995).
6. A. Pottier, C. Chanéac, E. Tronc, L. Mazerolles and J.-P. Jolivet, *J. Mater. Chem.* **11**, 1116 (2001).
7. J. H. Lee and Y. S. Yang, *Mater. Chem. Phys.*, **93**, 237 (2005).
8. Y. Zheng, E. Shi, Z. Chen, W. Li and X. Hu, *J. Mater. Chem.* **11,** 1547 (2001).
9. D. C. Hurum, A. G. Agrios, K. A. Gray, T. Rajh and M. C. Thurnauer, *J. Phys. Chem. B*, **107**, 4545 (2003).

Dye-sensitized solar cells using localized surface plasmon of gold and silver nanoparticles with comb-shaped block copolymer

Mikio Enomoto, Katsuhiko Taniguchi, and Manabu Ihara*

Research Center for Carbon Recycling Energy,
Tokyo Institute of Technology
2-12-1 Ookayama Meguro-ku, Tokyo, 152-8550, Japan
*E-mail:mihara@chem.titech.ac.jp

The conversion efficiency of dye-sensitized solar cells using 2 μm thickness of TiO_2 and N3 dye $(RuL_2(NCS)_2 \cdot 2H_2O$ (L=2,2'-bipyridyl-4,4'-dicarboxylic acid)) was improved from 1.5 to 2.4% by adding Ag nanoparticles covered with comb-shaped block copolymer (amine AgNP). Two methods of adding amine AgNP to porous TiO_2 films were used. Absorption enhancement of the dyes on quartz substrates by adding AuNP was investigated. Absorption enhancement of the dyes occurred when adding the heat-treated amine AuNP, and the prepared AuNP covered with 16-mercapto hexadecanoic acid (thiol AuNP). Absorption enhancement was higher for Black Dye (=BD) $(RuL'(NCS)_3 \cdot 3TBA$ (L = 2,2',2''-terpyridyl-4,4',4''-tricarboxylic acid; TBA=tetrabutyl ammonium)) than for N3. The dependences of the absorption enhancement on the treating conditions could be explained by the distance between BD and AuNPs. The main absorption enhancement peak for thiol AuNP was longer side of wavelength than that for amine AuNP.

Introduction

Dye-sensitized solar cells (DSSC) have the advantage that expensive production equipment using vacuum chambers is not needed. Therefore, fabrication costs for commercial solar cells are expected to be lower than for silicon solar cells. However, the DSSC conversion efficiency and stability are still relatively low, so that improving DSSCs is an area of active research. To increase conversion efficiency, we are studying the effect of using a localized surface plasmon of metal nanoparticles (metal NP).

An excited surface plasmon is the state that the electrons in a metal vibrate in a group corresponding to the vibration of the electromagnetic field of incident light on the metal. An electromagnetic field is formed by surface plasmon excitation, and the coupled interaction between the incident light and the plasmon causes resonance. This is called surface plasmon resonance. Because the electrons in metal NP vibrate intensively during surface plasmon resonance, a strong electromagnetic field is formed by the vibration. Moreover, considering continuity of electromagnetic fields, penetration of strong electromagnetic fields by plasmon occurs in the vicinity of the metal NP. This penetrated strong electromagnetic field is an optical near-field, which is a kind of evanescent wave that does not propagate, and is therefore referred to as a locally enhanced electromagnetic field by localized surface plasmon of metal NP. If dye exists near the metal NP, the absorption will increase beyond the sum of that for metal NP and dye alone (1~3). This is called absorption enhancement by localized surface plasmon. We previously reported that vacuum deposited AgNP on quartz substrates enhanced absorption of N3 dye by as much as 149:1, and we first proposed application of surface plasmon to DSSCs (4).

Much research has been done on surface plasmon. Research relating to Surface Enhanced Raman Spectroscopy (SERS) (5) and detailed research on surface plasmon has been done.

Enhancements of photoelectric current were reported for several types of photovoltaic cells by using strong electromagnetic fields of surface plasmon (6~12). Photocarrier injection directly from metal nanoparticles into titanium oxide (TiO_2) was also reported (13,14). Research for plasmon use in biosensors has also been done (15).

We also reported that the conversion efficiency of DSSCs is improved by a locally enhanced electromagnetic field created by using localized surface plasmon of AgNP covered with a thiol-type surface modulator (16).

Metal nanoparticles are several to tens of nm in diameter, and they are covered with a surface modulator to prevent aggregation. metal NPs are therefore stable in a colloidal solution. It is important that dye approaches nanoparticles for the absorption enhancement by localized surface plasmon to occur. The negative effects that occur by adding metal nanoparticles are a back charge transfer from TiO_2 to electrolyte and an obstruction of photo carrier injection from the dye into TiO_2, which must also be considered. If metal NP and dye are merely mixed, absorption enhancement and photoelectric conversion efficiency improvement will not necessarily occur. Therefore, in addition to identifying a suitable combination of materials, a suitable processing method for combining NP and dye must also be developed.

In the work described here, we first fabricated DSSCs having AgNP with comb-shaped block copolymer. The photoelectric conversion efficiency was improved by adding AgNP. Second, the absorption-enhancing effect of the dye by Au nanoparticles on a quartz substrate was explored with the goal of improving the efficiency of DSSCs by using Au nanoparticles. Whereas Ag may react with iodine in electrolytes and cause degradation of a cell, the usage of Au can prevent NPs reacting with electrolytes because of its chemical stability. If AuNP can be used with DSSCs, stable DSSCs with enhanced absorption caused by metal NP can be made.

Experiments

Two dyes used for DSSCs

Two dyes were used: N3 ($RuL_2(NCS)_2.2H_2O$ (L = 2,2'- bipyridyl- 4,4'- dicarboxylic acid)), which has two wavelength peaks at about 420 and 525 nm, and Black Dye (BD: $RuL'(NCS)_3$:3TBA (L = 2,2',2''''-terpyridyl-4,4',4''-tricarboxylic acid ; TBA = tetrabutyl ammonium)), which has three wavelength peaks at about 430, 570, and 640 nm.

Fabrication of DSSCs with N3 and amine AgNP

TiO_2 slurry was prepared in a planet-type ball mill from ethanol, water, acetyl acetone, 10x diluted TritonX-100, polyethylene glycol 500,000 (PEG), and P-25 TiO_2 powder with average particle diameter of 30 nm (AEROSIL).

A TiO_2 porous film was formed on the tin-oxide coated side of a 1x2 cm piece of glass (ASAHI GLASS, U film) by spin-coating the TiO_2 slurry described in the previous paragraph. The resulting film thickness was about 2 μm, as measured with a scanning electron microscope (SEM). Such thin films were used to clarify the effect of absorption enhancement by using localized surface plasmon. The film was sintered with the following cycle: 100°C for 5 minutes, followed by 350°C for 3 minutes, followed by 450°C for 20 minutes, and finally 350°C for 3 minutes. Because the sheet resistance increases rapidly if heated beyond 450°C, the temperature of the U film electrically conductive glass was kept below 450°C.

The TiO_2 film was shaved off from the substrate, wiped with ethanol, and trimmed to make 5x5 mm square film electrodes. Two kinds of amine AgNP (AgNP covered with comb-shaped block copolymer) supporting methods, subsequent dipping method and co-dipping method, were used. Subsequent dipping method is the way that TiO_2 films were dipped into dye solution after metal solution. Co-dipping method is the way that TiO_2 films were dipped into dye and metal mixed solution. In subsequent dipping methods, these electrodes were then dipped into an amine AgNP ethanol solution (NIPPON PAINT) with amine AgNP concentrations of $4x10^{-9}$, $5x10^{-8}$, and $1x10^{-7}$ [mol/l]. The electrodes were then dipped into a $3.0x10^{-4}$ [mol/l] N3 dye solution to support N3. TiO$_2$ film electrodes that supported only N3 were also made. In the co-dipping method, mixed solutions of amine AgNP and N3 were prepared with the following concentrations: (1) N3 (1.5×10^{-4} [mol/l]) only, (2) N3 (1.5×10^{-4} [mol/l]) + AgNP (2.5×10^{-8} [mol/l]), (3) N3 (1.5×10^{-4} [mol/l]) + amine AgNP (5×10^{-8} [mol/l]), and (4) N3 (1.5×10^{-4} [mol/l]) + amine AgNP (5×10^{-7} [mol/l]).

To make platinum counter electrodes, a conductive glass with two holes was made, into which electrolyte was poured. The electric conduction side was turned up, and the conductive glass was heated to 220°C using a hot plate. We then dropped a H$_2$PtCl$_6$·6H$_2$O propanol solution onto the conductive glass drop-by-drop so that a spot-like Pt film was deposited until the Pt film reached $1x10^{-6}$ mol/cm^2.

An iodine-type electrolyte was prepared as follows. 4-tert-Butylpyridine (Aldrich) : 0.5 [mol/l], lithium iodide (Wako 1st Grade, Wako Pure Chemical Industries) : 0.1 [mol/l], iodine (Special Grade, KANTO CHEMICAL) : 0.1 [mol/l], and dimethylpropylimidazolium iodide (SOLARONIX) : 0.6 [mol/l] were dissolved in 3-methoxypropionitrile.

The absorption spectrum of the TiO$_2$ film so prepared was then measured.

The platinum counter electrode was heated on a hot plate at 120°C, and the glasses were pushed together from the top to paste them together.

We then poured the electrolyte into the holes of the counter electrode. Nonconductive glass heated to 180°C was placed on the polymer resin and pasted to it. To reduce ohmic resistance, Ag paste was applied to the electrode conduction side.

Incident photo to current efficiency (IPCE) that describes the ratio of the carrier electron flux flowing in DSSC to the incident photon flux and I-V characteristic curves of the DSSC were measured, together with the absorption spectra of TiO$_2$ electrodes supported by amine AgNP and N3.

Measurement of composite film absorption spectra using amine AuNP on quartz substrates

To evaluate the absorption-enhancing effect, composite films covered with the comb-shaped block copolymer AuNP solution (NIPPON PAINT) and with the dye solution were deposited onto quartz substrates.

First, 10 μl of $3x10^{-1}$ wt% AuNP solution was dropped onto the quartz substrate and dried in air. The absorption spectra of these AuNP films were measured. Then, 10 μl of $5x10^{-4}$ [mol/l] N3 dye solution was dropped onto the films and dried in air. The absorption spectra of these composite films with AuNP and N3 were measured. By comparing the absorption spectra for each deposition step, the increased absorption due to the dye was evaluated.

Next, to evaluate the absorption-enhancing effect when the surface modulator of AuNP was removed, the AuNP film was heated at 310 °C for 3 minutes on a hot plate and cooled to room temperature. Then, either $3x10^{-1}$ wt% AuNP, $5x10^{-4}$ [mol/l] N3, or $1x10^{-3}$ [mol/l] BD dye solution was used to create a composite film, and the film absorption measured to determine which dye created the largest absorption enhancement.

To investigate how effectively the surface modulator was removed by heat treatment, TG-DTA thermoanalysis was applied to 7.5 mg dried AuNP by heating the sample from room temperature to 500°C at a rate of 3°C/min in the air. The TG-DTA data showed how much extent the surface modulators were oxidized and thermally decomposed at each temperature.

The effect of the concentration of AuNP and BD on the absorption-enhancing effect was also investigated. For a BD concentration of $1x10^{-3}$ [mol/l], the AuNP concentration was varied from $3x10^{-3}$ to $3x10^{-1}$ wt%. For an AuNP concentration of $3x10^{-1}$ wt%, the BD concentration was varied from $5x10^{-4}$ to $3x10^{-3}$ [mol/l]. Each composite film was heat treated according to the procedure described above and the absorption spectra measured. The absorption enhanced ratio due to surface plasmon was calculated in each composite film.

Preparation of AuNP with thiol-type surface modulator

To compare the property of comb-shaped block copolymer covered AuNP (amine AuNP) with another type of AuNP, we prepared AuNP covered with a 16-mercapto hexadecanoic acid used as a thiol-type surface modulator (thiol AuNP). If the absorption-enhancing effect caused by using localized surface plasmon is affected by the surface modulator, the conversion efficiency should also be affected. By comparing the absorption-enhancement effect of each AuNP, we evaluated the influence of plasmon absorption spectrum and surface modulators on absorption enhancement.

The preparation of Au nanoparticles was done by using reported method (17) where an $HAuCl_4 \cdot 3H_2O$ solution was mixed and stirred with a $(C_8H_{17})_4NBr$ chloroform solution for 1 hour. The chloroform phase was then removed from this solution with a separating funnel. 16-mercapto hexadecanoic acid was added to the chloroform solution and stirred for 15 minutes. $NaBH_4$ solution was then added and the solution stirred for 3 hours. Next, the chloroform phase was removed from the solution with a separating funnel.

We then put the chloroform solution and some amount of ion-exchanged water into a separating funnel, shook it to wash the chloroform phase, and again removed the chloroform phase. This operation was done twice to remove impurities, and finally the chloroform phase containing AuNP was used.

We then added ethanol and water to the chloroform solution, separated this solution by using a centrifuge, causing nanoparticles to be precipitated. After removing the supernatant liquid, ethanol and water were added, and again we did centrifugal separation. In this way, centrifugal separation was done 4 times. The precipitated thiol AuNP was dissolved by using ethanol. This solution is referred to as the undiluted thiol AuNP solution. After that, thiol AuNP solution was used by diluting with ethanol.

TG-DTA thermoanalysis was applied to 4.3 mg dried AuNP by heating the sample from room temperature to 500°C at a rate of 3°C/min in the air.

Measurement of composite film absorption spectra using thiol AuNP on quartz substrate

Thiol AuNP film properties were measured using similar methods as used to measure amine AuNP film properties. Composite films of thiol AuNP and BD were made without heat treatment, and with heat treatment at 310 °C for 3 min. For a BD concentration of $1x10^{-3}$ [mol/l], the thiol AuNP concentration was varied from diluted to 0.05 to undiluted. For undiluted thiol AuNP, the BD concentration was varied from $5x10^{-4}$ to $3x10^{-3}$ [mol/l]. Each composite film was heat treated and the absorption spectra measured. The absorption enhanced ratio due to BD in each composite film was calculated.

Results and Discussion

Characteristics of DSSC supported with amine AgNP and N3

For DSSCs using AgNP vacuum deposited on quartz substrates, an increase of N3 absorption by as much as 149:1 was achieved. The 550 nm wavelength band of N3 was the most enhanced, which is larger than the wavelength of peak absorption of 470 nm for AgNP deposited on quartz. Because locally enhanced electromagnetic field of metal NP have the same frequency as the plasmon absorption, absorption enhancement will occur for longer wavelengths, which has the lower energy, than surface plasmon absorption of metal NP.

Figure 1 show absorption spectra of photoabsorbing layer (TiO$_2$ + dye only or dye and amine AgNP) and (b) IPCE spectra, (c) I-V characteristic curve for DSSCs with N3 and with amine AgNP+N3 when using the subsequent dipping method. Absorption spectrum of amine AgNP solution $(1\times10^{-8}$ [mol/l]) used for the dipping is also plotted in Fig. 1(a).

Fig. 1(a) indicates that the absorbance of composite film at 520 nm increased with increasing amine AgNP concentration. This result and the enhancement of N3 absorption at about 550 nm in previous report (1) implied that the amine AgNP plasmon peak at 420 nm enhanced N3 absorption at 520 nm.

On the other hand, the DSSC efficiency did not show a corresponding increase. Figures 1(b) and 1(c) indicate that for an amine AgNP concentration of 4×10^{-9}[mol/l], the conversion efficiency successfully improved, from 1.5% for N3 $(3.0\times10^{-4}$ [mol/l]) only(1), up to 2.1% for a mixture of N3 $(3.0\times10^{-4}$ [mol/l]) and amine AgNP $(4\times10^{-9}$ [mol/l]) (2). However, for 5×10^{-8} (3) and 1×10^{-7} M (4) amine AgNP, there was no significant increase in efficiency compared with N3 only.

Figure 2 shows absorption spectra of photoabsorbing layer (TiO$_2$ + dye only or dye and amine AgNP) and IPCE spectra, I-V characteristic curve for DSSCs with N3 and with amine AgNP + N3 when using co-dipping method. In this case, the conversion efficiency was improved, from 1.5% for N3 $(1.5\times10^{-4}$ [mol/l]) only(1), up to 2.4% for a solution of N3 $(1.5\times10^{-4}$ [mol/l]) + amine AgNP $(5\times10^{-8}$ [mol/l]) (3).

TABLE I shows summary of the characteristic performance parameters of DSSCs.

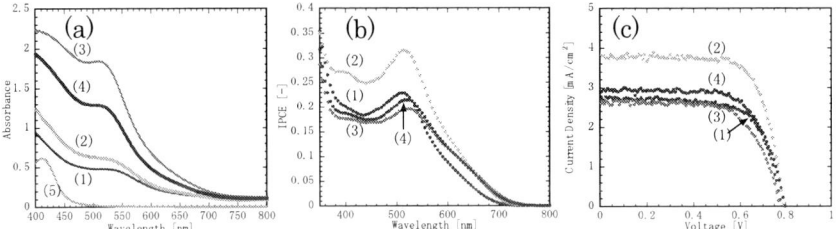

Figure 1. (a) Absorption spectra of photoabsorbing layer (TiO$_2$ + dye only or dye and amine AgNP) and (b) IPCE spectra, (c) I-V characteristic curve for DSSCs fabricated by subsequent dipping into solutions of (1) N3 $(3.0\times10^{-4}$ [mol/l]) only, (2) N3 $(3.0\times10^{-4}$ [mol/l]) + amine AgNP $(4\times10^{-9}$ [mol/l]), (3) N3 $(3.0\times10^{-4}$ [mol/l]) + amine AgNP $(5\times10^{-8}$ [mol/l]), and (4) N3 $(3.0\times10^{-4}$ [mol/l]) + amine AgNP $(1\times10^{-7}$ [mol/l])
(5) Absorption spectrum of amine AgNP solution $(1\times10^{-8}$ [mol/l]) used for the dipping.

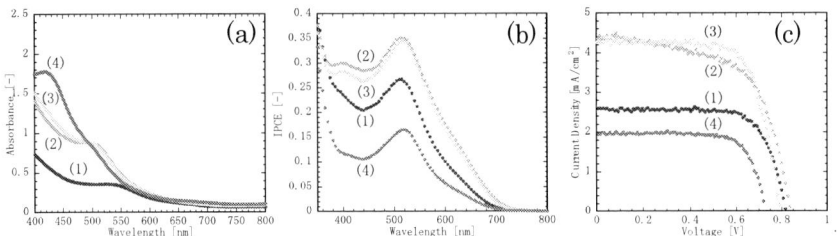

Figure 2. (a) Absorption spectra of photoabsorbing layer (TiO$_2$ + dye only or dye and amine AgNP) and (b) IPCE spectra, (c) I-V characteristic curve for DSSCs fabricated by co-dipping into solutions of (1) N3 $(1.5\times10^{-4}$ [mol/l]) only, (2) N3 $(1.5\times10^{-4}$ [mol/l]) + amine AgNP $(2.5\times10^{-8}$ [mol/l]), (3) N3 $(1.5\times10^{-4}$ [mol/l]) + amine AgNP $(5\times10^{-8}$ [mol/l]), and (4) N3 $(1.5\times10^{-4}$ [mol/l]) + amine AgNP $(5\times10^{-7}$ [mol/l])

For electric generation in DSSCs, the dye must directly contact TiO$_2$ and a transfer pathway must exist for the carrier electron excited by incident light. It is thought that metal NP with a modulator does not act as an electron transfer pathway. By having excess supported amine AgNP on TiO$_2$ surfaces, direct contact between N3 and TiO$_2$ decreases. Consequently, despite enhancement of the dye absorption and increased carrier-electron excitation, the transfer pathway for carrier electrons decreases. This apparently causes a decrease of the conversion efficiency.

In addition, by supporting metal NP on TiO$_2$, reverse electron transfer of a carrier electron into metal NP from TiO$_2$ will be promoted, causing the net conversion efficiency to decrease. Because these positive and negative factors are influenced by the concentrations of N3 and amine AgNP and by the dipping method, the conversion efficiencies of DSSC with amine AgNP varies with the concentration of N3 and amine AgNP and with the dipping method.

TABLE I. Characteristic performance of DSSCs in Figs. 1(c) and 2(c)

DSSCs fabricated by subsequent dipping (Fig. 1)	Voc [V]	Isc [mA/cm^2]	FF [-]	η[%]
(1) N3 (3.0×10^{-4} [mol/l]) (without Ag nanoparticles)	0.78	2.7	0.70	1.5
(2) N3 (3.0×10^{-4} [mol/l]) + amine AgNP (4×10^{-9} [mol/l])	0.79	3.8	0.71	2.1
(3) N3 (3.0×10^{-4} [mol/l]) + amine AgNP (5×10^{-8} [mol/l])	0.77	2.6	0.69	1.4
(4) N3 (3.0×10^{-4} [mol/l]) + amine AgNP (1×10^{-7} [mol/l])	0.80	2.9	0.70	1.6
DSSCs fabricated by co-dipping (Fig. 2)	Voc [V]	Isc [mA/cm^2]	FF [-]	η[%]
(1) N3 (1.5×10^{-4} [mol/l]) (without Ag nanoparticles)	0.82	2.6	0.72	1.5
(2) N3 (1.5×10^{-4} [mol/l]) + amine AgNP (2.5×10^{-8}[mol/l])	0.84	4.4	0.62	2.2
(3) N3 (1.5×10^{-4} [mol/l]) + amine AgNP (5×10^{-8} [mol/l])	0.79	4.3	0.72	2.4
(4) N3 (1.5×10^{-4} [mol/l]) + amine AgNP (5×10^{-7} [mol/l])	0.72	1.9	0.75	1.1

Composite film absorption spectra of amine AuNP on quartz substrate

Figure 3 shows comparisons of absorption peaks of amine AuNP, N3, and BD on quartz substrates. Figure 3 indicates that N3 peaks at 520 nm, on the short wavelength side of the peak of 550 nm for amine AuNP. On the other hand, BD has a large, gently-sloping absorption peak between 500~650 nm, on the long wavelength side of the peak of 550 nm for amine AuNP. Absorption enhancement by localized surface plasmon therefore occurs on the long wavelength side from the plasmon peak of metal NP. This implies that compared to BD, absorption can be enhanced more effectively by amine AuNP than by N3.

For absorption enhancement by localized

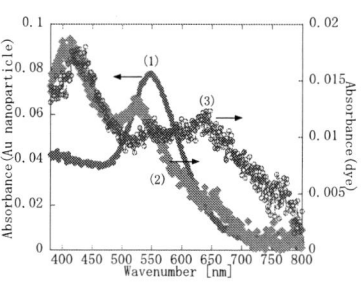

Figure 3. Absorption peaks for (1) amine AuNP, (2) N3 (3) BD deposited on quartz substrates.

surface plasmon, the distance between metal NP and dye is important. Because localized surface plasmon occur only in the metal NP neighborhood, dye must sufficiently approach metal NP to receive absorption enhancement due to the optical near field. However, the metal NP surface modulator may block the dye. When composite films with amine AuNP and N3 were made by simply dropping them onto the quartz substrate, absorption was not enhanced. To remove the surface modulator, which was suspected to block the dye, AuNP films dropped onto a quartz substrate were heated at 310 °C for 3 minutes, and then a composite film was made. Figure 4 indicates that thermoanalysis TG-DTA done on amine AuNP films removed about 37% of the surface modulator. The heat treatment allowed the dye to approach the AuNP, causing increased absorption by localized surface plasmon.

Composite films of AuNP heat-treated at 310 °C for 3 minutes and containing N3 or BD dye with comparable absorption were prepared, and the absorption spectra measured. Compared with the simple sum of heated AuNP and dye spectra only (4), composite film absorption (3) increased about 20% with N3 (Fig.5(a)) and 30% with BD (Fig.5(b)).

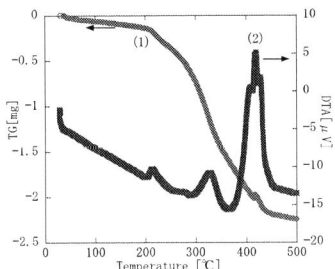

Figure 4. TG(1)-DTA(2) spectra for amine AuNP heat treated from 25 to 500 °C at a rate of 3 °C/min. Amine AuNP sample which was volatilized its solvent was 7.5 mg. 37% of the surface modulator was removed at 310 °C.

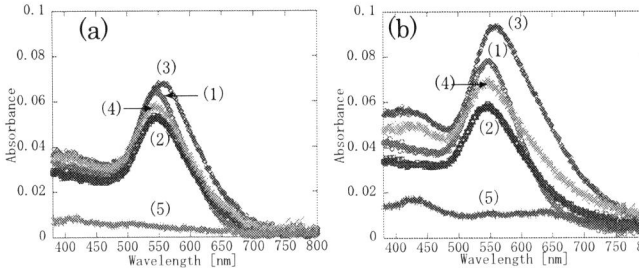

Figure 5. Absorption spectra of (1) 3×10^{-1} wt% AuNP, (2) heated AuNP, (3) composite film of AuNP and dye, (4) simple sum of heated AuNP and dye, (5) dye only.
Dye is (a) 5×10^{-4} [mol/l] N3, or (b) 1×10^{-3} [mol/l] BD.

This indicates that dye absorption enhancement occurs by heat treatment to amine AuNP, and the absorption enhancement by amine AuNP is stronger for BD than for N3. The effect of heat treatment confirms that the absorption enhancement by localized surface plasmon is influenced by the distance between metal NP and dye. The fact that absorption enhancement is stronger for BD suggests that the long wavelength side of the plasmon peak tends to cause the absorption enhancement. Therefore, BD was used as the dye in the following experiments.

To determine the level of absorption enhancement that AuNP gives to BD, composite films with varying concentrations of AuNP and BD were made, and the enhanced ratio measured. For this purpose, we define "enhanced BD absorption" and "enhanced ratio" as

"Enhanced BD abs." = "Composite film abs." $-$ "Heated AuNP film abs."
"Enhanced ratio" = "Enhanced BD abs." / "BD only film abs."

Figure 6 shows absorption spectra of composite films for (a) $3 \times 10^{-1} \sim 3 \times 10^{-3}$ wt% AuNP and 1×10^{-3} [mol/l] BD, and for (b) 3×10^{-1} wt% AuNP and $3 \times 10^{-3} \sim 5 \times 10^{-4}$ [mol/l] BD. Figure 7 shows the enhanced BD spectra for various amine AuNP solution concentrations (Fig. 7(a)) and various BD solution concentrations (Fig. 7(b)). Figure 7(a) indicates that absorption near 570 nm increased with increasing AuNP concentration. However, Fig. 7(b) indicates that for high BD concentration, enhancement occurred at both 570 and 650 nm. This indicates that the wavelength of the enhanced absorption is affected by the ratio of the concentration of AuNP and BD.

The BD absorption enhanced ratio at 570 and 650 nm based on Figs. 7(a) and 7(b) is shown in Figs. 8(a) and 8(b), respectively. The enhanced ratio increased faster at 570 nm than at 650 nm, and was proportional to the AuNP concentration at 570 nm. This is because 570 nm is closer than 650 nm to the AuNP plasmon peak that occurs between 520~530 nm. By contrast, when the BD concentration was increased, the absorption enhancement was not as high as the concentration increase, causing the overall enhanced ratio to decrease.

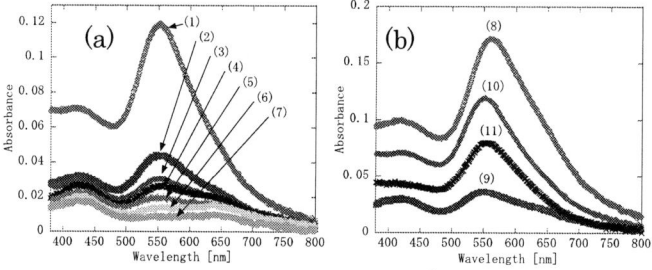

Figure 6. (a) Absorption spectra of composite films with 1×10^{-3} [mol/l] BD and various amine AuNP colloidal solutions ((1) 3×10^{-1}, (2) 1×10^{-1}, (3) 6×10^{-2}, (4) 3×10^{-2}, (5) 1×10^{-2}, and (6) 3×10^{-3} [wt%]), and for (7) 1×10^{-3} [mol/l] BD only.
(b) Absorption spectra of composite films with amine AuNP colloidal solutions of 3×10^{-1} [wt%] and BD solutions of (8) 3×10^{-3}, (9) 2×10^{-3}, (10) 1×10^{-3}, and (11) 5×10^{-4}[mol/l].

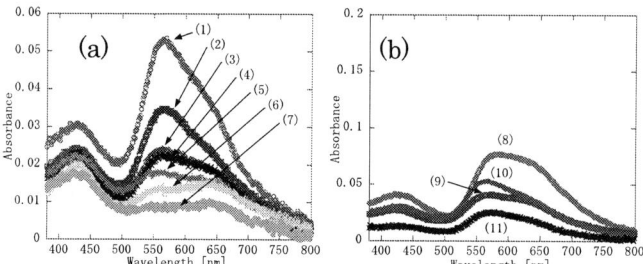

Figure 7. Enhanced BD absorption spectra of films shown in Fig.6.

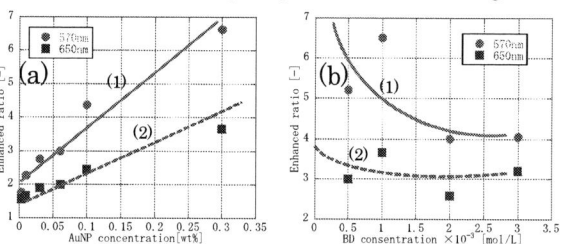

Figure 8. BD absorption enhancement ratio at (1) 570 and (2) 650 nm for various (a) amine AuNP solution concentrations, and for various (b) BD solution concentrations.

Figure 9 shows composite films for increased AuNP and BD concentration, which suggests that the distance between AuNP and BD is related to their concentrations. If AuNP concentration increases, more BD can approach AuNP, which increases the localized surface plasmon concentration. This in turn increases the enhanced ratio. Conversely, if BD increases, the concentration of BD that cannot approach AuNP increases, causing a decrease in the enhanced ratio.

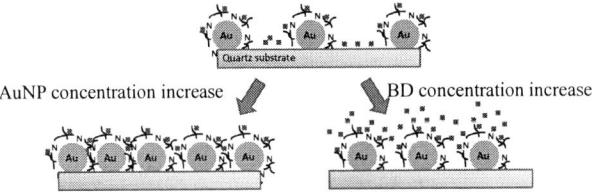

Figure 9. Composition of films for AuNP or BD concentration increases.

Preparation of AuNP with thiol type surface modulator

We synthesized thiol AuNP by changing the molar ratio between Au and thiol. Figure 10 shows the absorption spectrum of a thiol AuNP solution for a molar ratio of Au:thiol = 3:1, which indicates that plasmon absorption at 510 nm (peculiar to AuNP) was obtained. This solution is called the undiluted thiol AuNP solution.

Figure 11 shows the results of TG-DTA thermoanalysis done on thiol AuNP. According to Fig. 11, the surface modulator was removed starting at 200 °C, and at least 75% of the entire surface modulator was removed by 310 °C. Compared to amine AuNP, the surface states of thiol AuNP differed considerably, and the difference might cause the different absorption enhancement by localized surface plasmon after heat treatment.

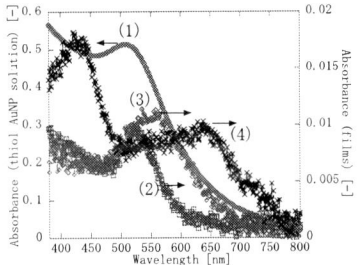

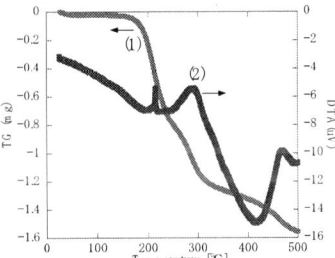

Figure 10. Absorption spectrum of (1) thiol AuNP colloidal solution, and absorption spectra of film for (2) amine AuNP, (3) thiol AuNP, and (4) BD on quartz substrate.

Figure 11. TG(1)-DTA(2) spectra for thiol AuNP heat treated from 25 to 500 °C at a rate of 3 °C/min. Amine AuNP sample which was volatilized its solvent was 4.3 mg. 75% of the surface modulator was removed at 310 °C.

Composite film absorption spectra using thiol AuNP on quartz substrate

Figure 12 shows absorption spectrum for (1) BD only, and enhanced BD absorption spectra for composite films of BD and thiol AuNP (2) without and (3) with heat treatment. Unlike amine AuNP, absorption enhancement occurred with and also without heat treatment. This is because for thiol AuNP, the strength of the bonding and stability between AuNP and surface modulator is sufficiently low that BD could approach thiol AuNP, even without heat treatment. Heat treatment further increased the absorption enhancement by removing the surface modulator.

Figure 10 shows a comparison of the absorption spectra for amine AuNP, thiol AuNP, and BD on a quartz substrate. Thiol AuNP had a longer plasmon peak than that of amine AuNP. Differences in the spectra indicate differences of the surface modulator state, differences in particle size and particle shape, etc. Thiol AuNP was expected to be more effective than amine AuNP in causing dye absorption enhancement by localized surface plasmon on the long wavelength side of the metal NP plasmon peak.

Similar to amine AuNP, the enhanced ratio of composite films of thiol AuNP and BD can be controlled by changing the film composition. Figure 13 shows absorption spectra of composite films for (a) thiol AuNP undiluted, diluted up to a ratio of 20:1 (0.05), and mixed with 1×10^{-3} [mol/l] BD; and for (b) undiluted thiol AuNP and mixed with $3 \times 10^{-3} \sim 5 \times 10^{-4}$ [mol/l] BD. Figure 14 shows the enhanced BD spectra for each films in Fig.13.

Figure 14(a) indicates that for the increase of thiol AuNP concentration, enhanced BD absorption increases smoothly above a wavelength of 550 nm. However, Fig.14(b) indicates that for the increase of BD concentration, enhanced BD absorption increased near 650 nm rather than at 570 nm. In each film, thiol AuNP enhanced longer wavelength absorption than for amine AuNP. This result supports our previous expectation that thiol AuNP would enhance the longer wavelength side more than amine AuNP.

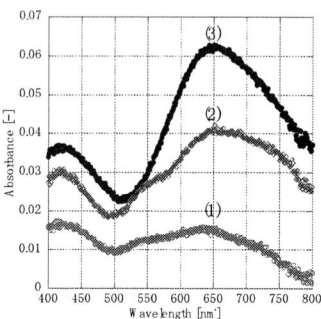

Figure 12. Absorption spectrum for (1) BD only, and enhanced BD absorption spectra for composite films of BD and thiol AuNP (2) without and (3) with heat treatment.

The enhanced ratio of BD absorption based on Fig. 14 at 570 and 650 nm is shown in Figs. 15(a) and 15(b), respectively. The value of amine AuNP is also shown in Figs. 15(a) and 15(b). Unlike amine AuNP, concentration of thiol AuNP was not measured in this experiment. To compare

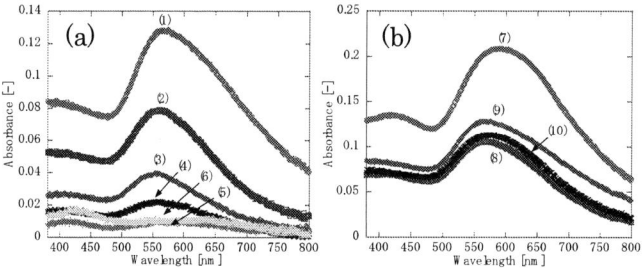

Figure 13. (a) Absorption spectra of composite films with BD concentration of 1×10^{-3} [mol/l] and various thiol AuNP colloidal solutions: (1) undiluted; diluted to (2) 0.5, (3) 0.3, (4) 0.15, and (5) 0.05; and spectrum of films with (6) BD concentration of 1×10^{-3} [mol/l] only. (b) Absorption spectra of composite films with undiluted thiol AuNP colloidal solutions and BD concentrations of (7) 3×10^{-3}, (8) 2×10^{-3}, (9) 1×10^{-3}, and (10) 5×10^{-4} [mol/l].

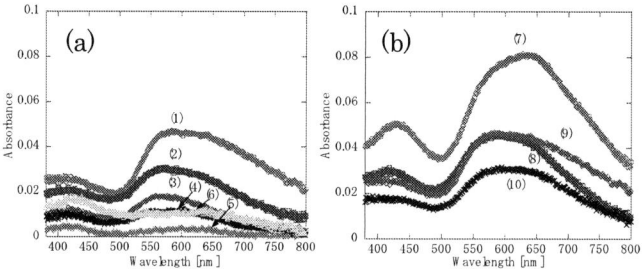

Figure 14. Enhanced BD absorption spectra of each film shown in Fig.13.

with each BD enhanced ratio for thiol AuNP and amine AuNP, these data were plotted by the peak absorption values of AuNP films on a quartz substrate instead of the concentrations in Fig. 15(a). Figure 15(a) shows that almost all enhanced ratios for thiol AuNP were less than for amine AuNP. However, when the AuNP peak absorption was as high as shown in Fig. 15(b), the enhanced ratio for thiol AuNP at 650 nm increased much than for amine AuNP. This is because the plasmon peak of thiol AuNP occurred at a longer wavelength than for amine AuNP.

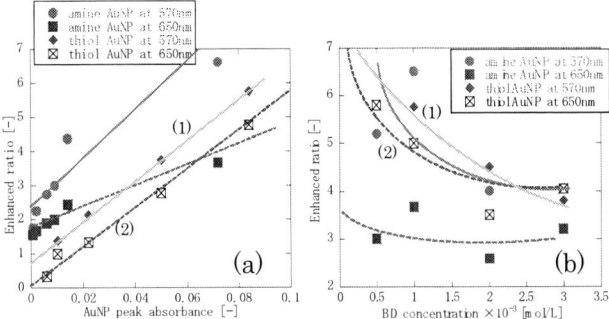

Figure 15. BD enhancement ratio at (1) 570 and (2) 650 nm for various (a) thiol AuNP solution concentrations and for various (b) BD solution concentrations.

Similar to amine AuNP, overall the enhanced ratio tended to decrease for BD concentrations approaching those of AuNP. It is thought that the change in the enhanced ratio is related to the distance between AuNP and BD, which is related to their concentrations.

Conclusions

Compared to DSSCs using N3 dye, which has a conversion efficiency of 1.5%, DSSCs made co-dipping with N3 and amine AgNP covered with comb-shaped block copolymer has a conversion efficiency up to 2.4%. However, for amine AgNP solution concentrations greater than 5×10^{-8} [mol/l], even though the absorption increased compared to N3 supported on DSSC films, the conversion efficiency did not increase. Possible reasons are that for amine AgNP supported on TiO_2 films, the amount of N3 in direct contact with TiO_2 will decrease and the back charge electron transfer from TiO_2 into electrolyte will increase. This implies that the carrier electron transfer from N3 to TiO_2 also decreased.

Comb-shaped block copolymer covered AuNP (amine AuNP) causes absorption enhancement for the dye, but only when the film was heat treated. However, thiol AuNP films had enhanced absorption even without heat treatment. This is because the localized surface plasmon affinity to dye differed between the amine and thiol AuNP, due to differences in the bonding strength of the surface modulator to AuNP and due to differences in AuNP particle diameter.

BD absorption was enhanced by amine AuNP more than N3, and the enhancement was the strongest at 570 nm. Because thiol AuNP had a higher plasmon peak on the long wavelength side than amine AuNP, the enhanced ratio of thiol AuNP was 570 nm, lower than for amine AuNP, and the enhanced ratio increased in the range of 600~700 nm.

Increasing the BD concentration with keeping AuNP concentration constant caused the enhanced ratio to decrease, because the BD that could receive localized surface plasmon of AuNP decreased. This effect occurs in two kind of surface modulator. To effectively enhance dye absorption by metal NP, appropriate concentration ratio of metal NP to dye used in the film is required.

References

(1) Joseph R. Lakowicz, *Analytical Biochemistry*, **337**, 171-194 (2005)

(2) T. Yatsui, Y. Nakajima, W. Nomura, and M. Ohtsu, *Applied Physics B: Lasers and Optics*, **84**, 265-267 (2006)

(3) Glass, A. M.; Liao, P. F.; Bergman, J. G.; Olson, D. H. *Opt. Lett.* **5**, 368, (1980)

(4) Manabu Ihara, Kanako Tanaka, Kenji Sakaki, Itaru Honma, Koichi Yamada, *J. Phys. Chem. B* **101**, 5153, (1997).

(5) Libin Yang, Xin Jiang, Weidong Ruan, Jingxiu Yang, Bing Zhao, Weiqing Xu, and John R. Lombardi, *J. Phys. Chem. C* **113**, 16226–16231, (2009)

(6) Min Wei Chen, Yuan-Fong Chau, Din Ping Tsai, *Plasmonics,* **3**, 157-164 (2008)

(7) K. Yamada, K. Miyajima, F. Mafune, *J. Phys. Chem. C*, **111**, 11246-11251 (2007)

(8) T. Shimada, S. Tomita et al, *Japanese Journal of Applied Physics* **48**, 042001 (2009)

(9) T. Arakawa, T. Muraoka, T. Akiyama, S. Yamada, *J. Phys. Chem. C* **113**, 11830–11835, (2009)

(10) K. Sugawa, T. Kawahara, T. Akiyama, S. Yamada, *Japanese Journal of Applied Physics* **48**, 04C132, (2009)

(11) Wei Hao Lai, Yen Hsun Sub, Lay Gaik Teoh, Min Hsiung Hon, *Journal of Photochemistry and Photobiology A: Chemistry,* **195**, 307–313, (2008)

(12) Z. H. Chen, Y. B. Tang, C. P. Liu, Y. H. Leung, G. D. Yuan, L. M. Chen, Y. Q. Wang, I. Bello, J. A. Zapien, W. J. Zhang, C. S. Lee, and S. T. Lee, *J. Phys. Chem. C* **113**, 13433–13437, (2009)

(13) Y. Takahashi, T. Tatsuma, *Electrochemistry Communications,* **10**, 1404–1407 (2008)

(14) A. Zhu, Y. Luo, Y. Tian, *Anal. Chem.* **81**, 7243–7247 (2009)

(15) Y. Sun, D. Song, Y. Bai, L. Wang, Y. Tian, H. Zhang, *analytica chimica acta*, **624**, 294–300 (2008)

(16) Rihito Ito, Kichiro Oryu, Hideshi Shibuya Manabu Ihara, *PVSEC-17 extended abstract*, 6P-P5-05. (2007)

(17) Mathias Brust, Merry1 Walker, Donald Bethell, David J. Schiffrin and Robin Whyman J. *CHEM. SOC., CHEM. COMMUN.,* 801-802 (1994)

Photoelectrochemistry of Hematite Thin Films

Heli Wang and John A. Turner
National Renewable Energy Laboratory
1617 Cole Boulevard, Golden CO 80401, USA

Optical investigation indicated that both nanorod and pyrolysis hematite thin films show direct and indirect band gap transitions. Nanorod film shows a direct band gap of 2.14 eV and an indirect band gap of 2.04 eV. Pyrolysis film gives a direct gap of 2.15 eV and an indirect gap of 2.08 eV.

Pyrolysis film has much higher *IPCEs* than the nanorod film. *IPCE* measurements give band gaps of 1.94 and 2.00 eV for nanorod film and pyrolysis film, respectively. Photocurrent onsets in 1M NaOH solution were –0.44 V for nanorod film and –0.14 V for pyrolysis film. For nanorod film, linear (photocurrent)$^{0.5}$ ~ light intensity relationship means a high charge carriers' recombination. Linear photocurrent ~ light intensity relationship with pyrolysis film indicates a fast charge transfer. Adopting pyrolysis film in the dual photoelectrode assembly resulted in zero short circuit photocurrent, due to the mis-match of the CBM of the pyrolysis film with the VBM of the *p*-GaInP$_2$. It is expected that combining the advantages of the two films should form an efficient electrode for photoelectrochemical water splitting with efficient charge transfer and a large surface area.

Introduction

The photoelectrochemical (PEC) production of hydrogen and oxygen from water uses energetic electrons and holes, generated by absorbed photons, to perform water splitting reactions. This process was first demonstrated by Fujishima and Honda at a chemically biased TiO$_2$ with ultraviolet (UV) light (1). The efficiency of the process is largely determined by the photosensitivity of the photoelectrode. To explore an efficient and economical PEC process for renewable hydrogen production, significant effort has been focused on the semiconductor (SC) materials.

Compared to other materials, iron oxide (hematite) has several advantages as a semiconductor material for this purpose. With a band gap around 2 eV (2,3), it could utilize 40% of the incident solar spectra. It has an excellent chemical stability in a broad pH range (3) and its valence band is appropriate for oxygen evolution (2-4). Moreover, it is abundant on earth, low in cost, and non-toxic making hematite an attractive candidate for PEC water splitting. However, iron oxide has shown poor charge transport due to high recombination losses (4,5). It is generally accepted that recombination of photo-generated electrons and holes, trapping of electrons by oxygen deficiency sites, and low mobility of the holes causing a low conductivity are reasons of low photoresponse for iron oxide (2,4,5). Different approaches, including different doping (2,6-8) have been attempted to increase the photoresponse of iron oxide. However, very limited improvement is obtained with the bulk hematite pellets.

The other approach was using thin films. Miller and co-workers have successfully applied a thin Fe_2O_3 film as the top active layer of a hybrid multi-junction cell for hydrogen production (9,10). Significant improvement in photocurrent was obtained since the introduction of nanostructured hematite particle thin films (11-14). Because the traveling distance for the photogenerated holes is reduced with nanostructured spherical particles, the charge transport process is improved significantly. Nanocrystalline hematite thin films based on oriented nanorod arrays (15-17) further enhance the photogenerated electron transport along the nanorods due to fewer grain boundaries (possible recombination centers) and a directed electron movement toward the back contact. Other thin films of nanostructured hematite have been made and shown promising results (3,18,19). In particular, thin films prepared by spray-pyrolysis have shown high photoresponse (20-22). In this report, we will focus on the photoelectrochemical behavior of hematite thin films based on nanorod arrays and prepared by the spray-pyrolysis method.

Experimental

Materials and Characterization

Hematite nanorod thin film (nanorod film) was provided by the Department of Physical Chemistry at Uppsala University, Sweden. The detailed preparation procedure is available elsewhere (15-17). Typical thickness of such thin film is *ca.* 0.7 μm. The spray-pyrolysis hematite thin film (pyrolysis film) was provided by the Department of Chemistry at University of Geneva, Switzerland, and the preparation method was described in ref. 21. Pyrolysis film is 5% Ti-doped and is typical *ca.* 0.8~1 μm thick. Both films were deposited on a conducting glass (Tec 8) substrate.

The structures of the thin films were investigated by means of the glancing angle X-ray diffraction (XRD) pattern technique. This was conducted by means of a four-circle Scintag X-1 diffractometer (ThermoLab) with a Cu Kα anode source. To reduce the noise-to-signal ratio in a reasonable test period, a slow scanning speed of 0.01 degree/min and a preset time of 3 s/step were applied.

Optical measurements were carried out by means of an N&K Analyzer 1280 (N&K Technology, Inc.). In the measurements, both transmittance and reflectance were collected separately. A correction for scattering and transmittance from the conducting glass was made by assuming the validity of the equation (15):

$$\alpha d = -\ln\left[\frac{T_{H+S}/T_S}{1-R_{H+S}}\right] \qquad [1]$$

where α is the absorption coefficient, d is the film thickness, T_{H+S} and T_S are the transmittances of the entire film (hematite + substrate) and substrate alone, respectively, and R_{H+S} is the reflectance from the film. The reflectance from the substrate was found to be negligible.

Electrochemistry and Photoelectrochemistry

To make the (photo)electrodes, a thin film sheet was cut into smaller pieces and the substrate conducting glass was connected to a copper wire by means of conductive silver paint. Then, the edges and the place for electrical conduction were sealed with an epoxy resin a few times to eliminate possible leak.

1 M NaOH aqueous solution was prepared with reagent chemical and de-ionized water (resistance greater than 18 MΩ·cm). All (photo)electrochemical experiments were carried out at room temperature. A Pyrex™ glass cell with a volume of *ca.* 100~120 ml was employed as the test cell for (photo)electrochemical test. The cell was placed inside a dark box to avoid the influence of ambient light. A conventional 3-electrode configuration, with a platinum sheet as the counter electrode and a Ag/AgCl electrode (SSE) as the reference electrode was utilized for the measurements. So, the potentials in this paper refer to SSE except as otherwise specified.

Photoresponse measurements were carried out by means of a potentiostat (Solartron 1287) interfaced with a PC. In such tests, the electrode was first stabilized under light intensity of approximately 0.5 W/cm² (5 suns) at 0.7 V for 2 minutes. Then a potential scan was carried out towards the cathodic direction with chopped light, in which the light was chopped with 2 minutes ON and 2 minutes OFF alternatively. The scanning rate was 0.5 mV/s. A 250-W Oriel tungsten-halogen quartz lamp (Model 66183, Oriel Corporation) with a 6286A DC power supply (Hewlett Packard) was employed as the light source. According to the manufacturer's specification for the lamp, light intensity below 350 nm is so little that the output from the lamp can be treated as white (visible) light only. The light intensity was calibrated using an Astral AC2500 calorimeter and AD30 Power & Energy Measurement System (Scientech, Inc.). Moreover, to eliminate the heating up of the solution due to the *infra*red (IR) irradiation, a 55 mm diameter water filter (Newport) was placed in the beam routine between the light source and the cell.

The same setup was used to determine the dependency of photocurrent on the light intensity. In these experiments, the electrode was held at 0.5 V for a few minutes, selected based on the trial polarizations with chopped light, to stabilize in the dark. Then the electrode was illuminated at a certain light intensity for 2 minutes followed by 2 minutes in the dark again. This procedure was repeated with changing light intensities.

The photoelectrochemical setup for action spectra measurements consisted of a 100 W tungsten arc lamp housing with a LPS-220 lamp power supply (Photon Technology International), a Model SID-101 monochromator (Photon Technology International), a EG&G Model 263A Potentiostat/Galvanostat (Princeton Applied Research). In these tests, the iron oxide electrode was held at 0.5 V and the photocurrent at each monochromic light was registered. The incident photon-to-current efficiency (*IPCE*) was calculated according to (23):

$$IPCE_\lambda = \frac{1240 \cdot I_\lambda}{P_\lambda \cdot \lambda} \qquad [2]$$

where I_λ is the photocurrent in $\mu A/cm^2$, P_λ is the light power intensity in $\mu W/cm^2$ and λ is the wavelength in nm.

Results and Discussion

XRD Characterization

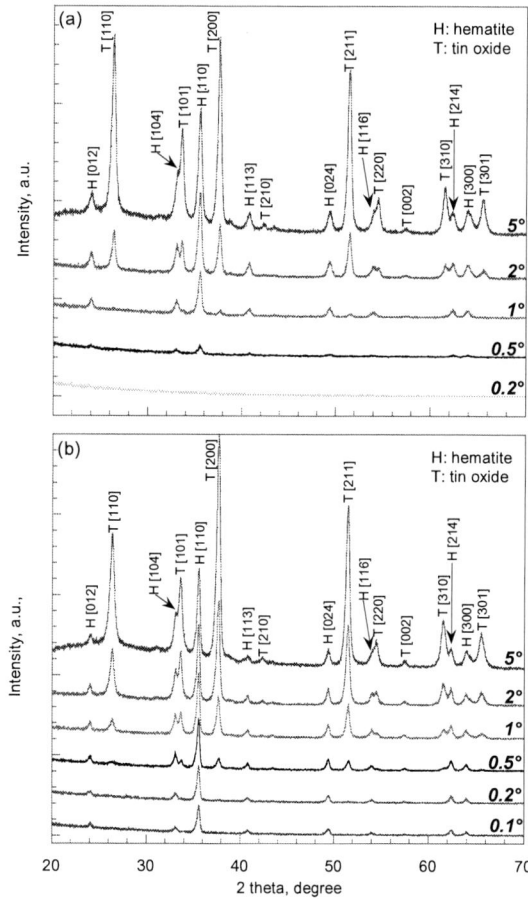

Figure 1. Grazing angle XRD patterns of the hematite nanorod thin film (a) and pyrolysis thin film (b). The grazing angles are marked on the right of the patterns.

A grazing angle XRD pattern of the nanorod thin film is shown in Figure 1a. Strong tin oxide peaks (marked with T) are seen in the pattern at grazing angle of 5°, which represents the bulk layer structure of the conductive tin oxide. Hematite peaks (marked with H) are also registered at this angle, and some are very strong (like H[110]), suggesting the crystallization orientation. At 2°, hematite peaks dominate the pattern, while the peaks of tin oxide are significantly reduced. At 1°, it seems that the peaks of tin oxide could be almost ignored. Only very small peaks of T[200] and T[211] could be seen, indicating that the tin oxide content is low at this angle. At 0.5°, however, all of the remaining peaks are hematite, indicating that the pattern at this angle represents only the hematite thin film. Further decreasing the grazing angle to 0.2°, the signal is so small that no peak could be registered practically. Figure 1a suggests that the nanorod thin film is pure hematite.

Figure 1b shows the gracing angle XRD pattern of a pyrolysis thin film. Carefully inspecting Figure 1b, it is noticed that tin oxide peaks are present until at grazing angle of 0.5°. No tin oxide peaks can be seen by further decrease the grazing angle. Moreover, hematite peaks are still rather strong even at the lowest available grazing angle of 0.1°. This illustrates that the pyrolysis film is denser (or more compact) than the nanorod film. This is reasonable since the nanorod film is composed of elongated nanocrystals with bundles of finer fibers and oriented perpendicularly to the substrate (15,17). However, it is difficult to identify the Fe_3O_4 peak(s) by means of XRD technique as mentioned by Sartoretti et al (21). In that paper, the Raman spectrum was used to identify the substance. From our XRD results, however, hematite is the predominant phase in the pyrolysis film.

Optical Band Gaps

It is well established that the following function gives a satisfactory description of the absorption behavior in this wavelength domain (24):

$$\alpha h \nu = C_1 (h \nu - E_g)^n \tag{3}$$

where α is the absorption coefficientgth, h is the Plank constant, ν is the frequency of the photon, C_1 is a constant, E_g is the band gap of the semiconductor, and the prefix n is 0.5 for a direct and 2 for an indirect band gap transition. Thus, the optical band gap can be analyzed by plotting the $(\alpha h \nu)^n$ (where $n=0.5$ or 2) against the input photon energy $(h \nu)$. Such plots are shown in Figure 2. Both direct and indirect band gaps are found for nanorod film, Figure 2a. They are 2.14 eV and 2.04 eV, respectively, which are in excellent agreement with previous observation (15). It is interesting to note that similar behavior is registered with pyrolysis film, Figure 2b. The direct and indirect band gaps for pyrolysis film are 2.15 eV and 2.08 eV, respectively. The band gap of nanorod film is smaller than that of pyrolysis film. This agrees with the visual inspection that nanorod film looks darker than the pyrolysis film though the latter is slightly thicker. Theoretically, smaller band gap will allow the nanorod film to absorb more light in the visible spectrum. However, high absorption does not directly result in high photocurrent, due to the low conductivity and high recombination loss of hematite. This will be discussed later.

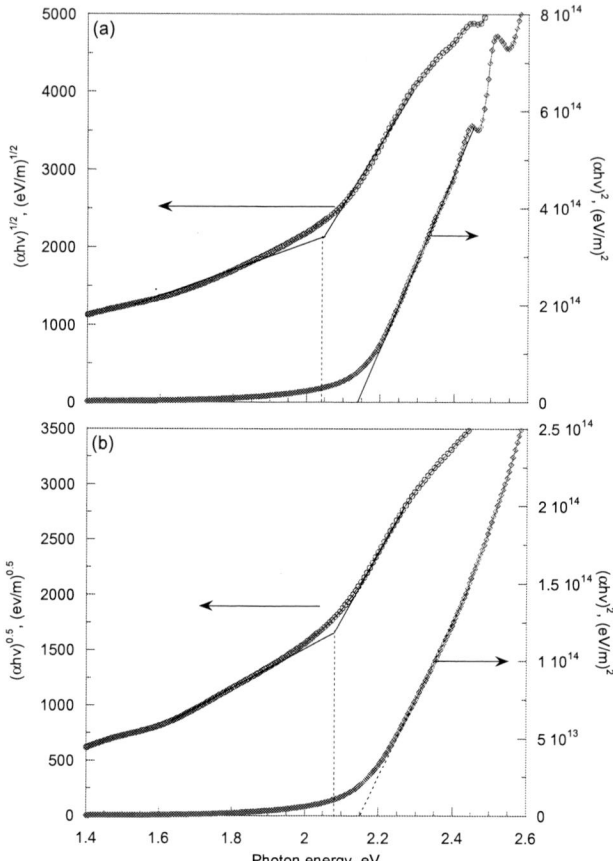

Figure 2. Indirect and direct band gaps of the hematite thin films. (a) Nanorod hematite film; (b) Pyrolysis hematite film.

Action Spectra

For photons with a wavelength λ, photon absorption rate A_λ that generates photocurrent via a Schottky junction in an n-type semiconductor is given by (25):

$$A_\lambda = P_\lambda \frac{1 - e^{-\alpha h \nu x_d}}{1 + \alpha L_p} \qquad [4]$$

in which P_λ is the photon flux intensity for photons with wavelength λ, x_d is the width of the depletion region and L_p is the diffusion length for holes.

Applying Equation [3] to [4], we have:

$$A_\lambda = P_\lambda \frac{1 - e^{-C_1 x_d (h\nu - E_g)^n}}{1 + \alpha L_p} \qquad [5]$$

The value of α increases with increasing photon energy. For photons with energy near the band gap, α becomes very small. If L_p is also small, which is the case for iron oxide (4,5), $\alpha L_p \ll 1$ thus making $1 + \alpha L_p \approx 1$. Moreover, with photon energy near the band gap, the argument of the exponential is small and can be replaced by the first two terms of its Taylor expansion and yield. Then, the above equation is simplified as:

$$A_\lambda = P_\lambda C_1 x_d (h\nu - E_g)^n \qquad [6]$$

Now, if the photocurrent is proportional to the photons absorption just described, i.e., $I_\lambda \propto A_\lambda$, the following equation can be used to describe the photocurrent as:

$$\frac{I_\lambda}{P_\lambda} = C_2 (h\nu - E_g)^n \qquad [7]$$

in which constant C_2 contains factors C_1 and x_d (the depletion width is constant at a constant potential) and the ratio within $I_\lambda \propto A_\lambda$. Combining Equations [2] and [7] will give:

$$IPCE_\lambda = \frac{1240 \cdot C_2}{\lambda} (h\nu - E_g)^n \qquad [8]$$

From Equation [8], plotting *IPCE* against the photon energy ($h\nu$) will then give the band gap of the thin films. Such *IPCE* plot for hematite thin films at 0.5 V is shown in Figure 3. *IPCE* values for nanorod film are very low, generally below 2%. A long tail at longer wavelength is registered for this film. These features agree well with the previous investigation of the hematite nanorod film (15,16). With pyrolysis film, *IPCE* values are generally rather high, in agreement with the previous reference (3). At 450 nm, *IPCE* of pyrolysis film is *ca.* 6 times higher than that of nanorod film. The *IPCE* of the pyrolysis film decreases almost linearly with the wavelength. At longer wavelength near the absorption edge, however, nanorod film shows higher *IPCE* than that of pyrolysis film (Inset of Figure 3), even though both values are not high. In general, the high *IPCE* with pyrolysis film indicates that this film is much more efficient in charge transfer. So, much better photoresponse and photocurrent would be expected with this film. On the contrary, lower *IPCE* values with nanorod film suggest that low charge carrier mobility and recombination of charge carriers could be the cause (15,16). Thus, lower photocurrent and photoresponse would be expected with this film. From Figure 3, the *IPCE* edges are *ca.* 640 nm for nanorod film and *ca.* 620 nm for pyrolysis film. These give corresponding band gap of 1.94 eV and 2.00 eV, respectively. Both agree well with the indirect optical band gaps.

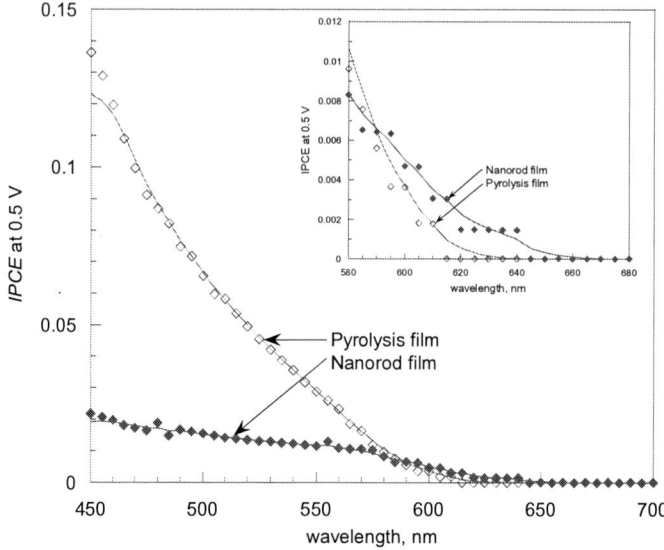

Figure 3. IPCE of the hematite thin films at 0.5 V in 1 M NaOH. Inset shows the details at light with longer wavelength.

Behavior under Chopped Light

Dynamic polarization with chopped light can generate both photocurrent and dark current in a single measurement. Such dynamic polarization for hematite thin films in 1 M NaOH solution is shown in Figure 4. Both films show n-type character in most of the potential range investigated. Low photocurrent and spikes are registered for the nanorod film. This agrees well with the action spectra of the film and again could be related to the recombination of the photo-generated charge carriers (16). Much higher photocurrent is registered for pyrolysis film, in agreement with the *IPCE* measurements, Figure 3. By carefully inspecting Figure 4, it is noticed that cathodic photocurrent is obtained in the cathodic potential region. Similar behavior was recorded previously with Nb-doped iron oxide (26). This suggests that hematite thin films show p-type character in the cathodic potential region. The reason for this is not fully understood. If we use the joint of n-type and p-type behavior as the photocurrent onset point, we obtain –0.44 V for nanorod film and –0.14 V for pyrolysis film, respectively. Both onsets agree well with previous investigations (15,16,22). The *ca.* 0.30 V anodic shift of the photocurrent onset for the pyrolysis film could be related to the magnetite (Fe_3O_4) identified by the Raman spectroscopy (21). Although a more anodic photocurrent onset is registered, much higher anodic photocurrent is seen with the pyrolysis film than that with the nanorod film. This is solely due to the efficient charge transfer process with the former.

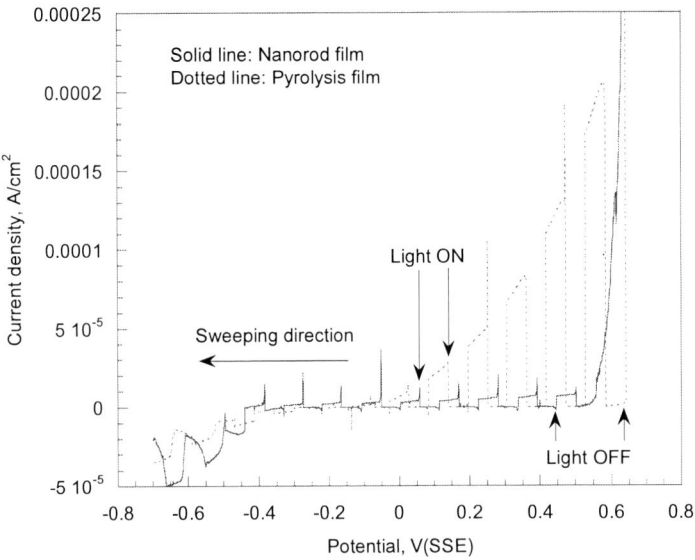

Figure 4. Dynamic polarization of thin hematite films in 1 M NaOH with chopped light. The potential sweeps to the cathodic direction, with a scanning rate of 0.5 mV/s. The light was chopped with 2 min ON and 2 min OFF alternatively. The light intensity was *ca.* 0.5 W/cm^2.

Effect of Light Intensity

The photocurrent of nanorod hematite film at 0.5 V at different light intensities up to 0.51 W/cm^2 is shown in Figure 5. As soon as the potential is applied in the dark, transient current decays very fast and stabilized. Dark current is comparable to the photocurrent level, especially at lower light intensities. Photocurrent increases with light intensity, though it is still in a few µA/cm^2 level even at light intensity of 0.51 W/cm^2. When illuminated at lower light intensities, the current spikes are smaller compared to those at higher light intensities. So by deducting for the dark current, we plotted the square root of net photocurrent against light intensity and shown in the inset of Figure 5. Linearity between square root of photocurrent and the light intensity is seen.

It has been well established (27,28) that linear photocurrent~light intensity relationship represents a low charge recombination and a fast charge transfer process in a semiconductor electrode/electrolyte interface; while a slow charge transfer process with semiconductor electrode would give a linear relationship of square root of photocurrent~ light intensity. The above results indicate that a high charge carrier recombination and slow charge transport process is predominant for the nanorod film. This agrees well with the action spectra and the polarization under chopped light, Figures 3 and 4. Modification of the material is needed to have an efficient photoelectrochemical water splitting process.

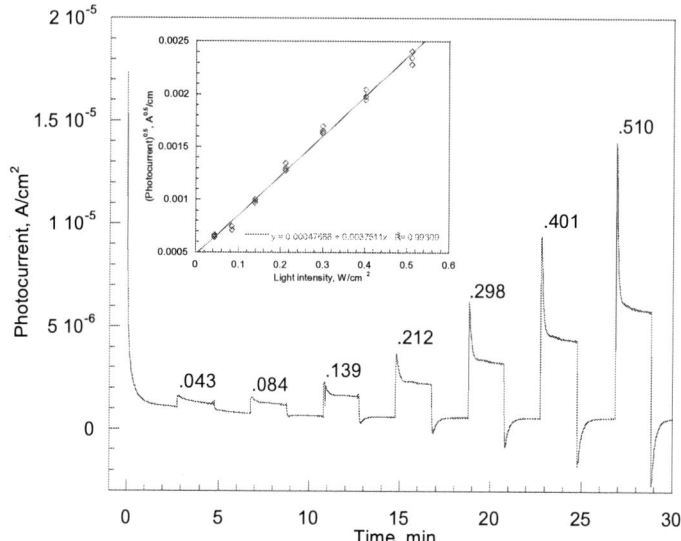

Figure 5. Influence of light intensity (marked in the plot) on the photocurrent of Nanorod hematite film at 0.5 V in 1 M NaOH solution. The light was shut off for 2 minutes after 2 minutes illumination. Inset shows the (photocurrent)$^{0.5}$ ~ light intensity relationship.

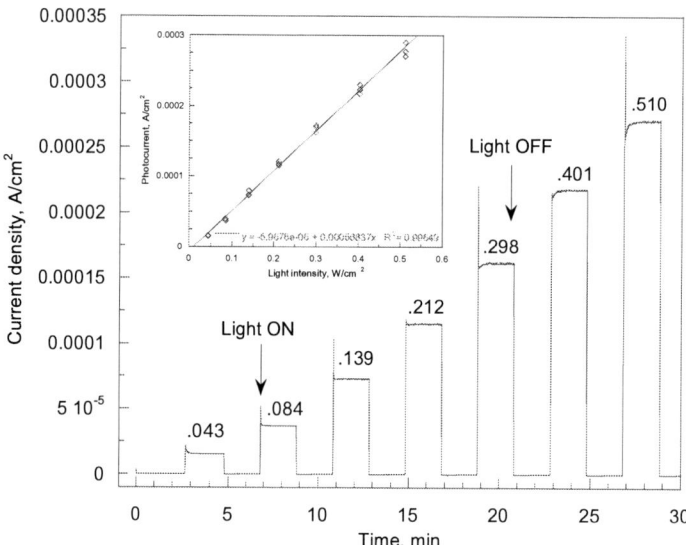

Figure 6. Influence of light intensity (marked on the plot) on the photocurrent of Pyrolysis film at 0.5 V in 1 M NaOH solution. The light was shut off for 2 minutes after 2 minutes illumination. Inset shows the linear relationship between photocurrent and light intensity. Note that the square of the relation co-efficient is better than 0.996.

Similarly, the photocurrent of pyrolysis film at 0.5 V at different light intensities up to 0.51 W/cm^2 is shown in Figure 6. Much higher photocurrents were obtained with the pyrolysis film. The dark current at this potential is so little that it can be ignored compared to the photocurrent level. Photocurrent spikes are seen when the electrode is illuminated, and it seems that this spike increases with the light intensity. Unlike the slow decay of photocurrent in the case of the nanorod film, the photocurrent of the pyrolysis film stabilizes very fast. This is due to the fast charge transport process with the pyrolysis film. Moreover, photocurrent increases with light intensity. Photocurrent against the light intensity plot is displayed in the inset of Figure 6. A linear photocurrent ~ light intensity relationship is obtained from the plot. The excellent linearity between photocurrent and light intensity clearly illustrates that the charge recombination is low and charge transfer process is fast at the pyrolysis hematite film electrode/electrolyte interface. The extrapolated line does not go through the origin indicating that some minimum light is required to generate a detectable photocurrent, as in the case of WO$_3$ thin films (29). These results agree well with the *IPCE* measurements, Figure 3.

It is well established that an efficient photoelectrochemical process is determined by two opposing factors: enough carriers must be present so that ohmic losses caused by the intrinsic resistivity of the semiconductor are minimized, yet the number of carriers present must be small enough so that the space charge layer produced by the semiconductor/electrolyte junction can have a significant extension into the bulk of the semiconductor (30). In this context, the combination of the advantages of the two films, i.e., the huge true surface area of nanorod film and the fast charge transfer with pyrolysis film would result in a better material for the application in photoelectrochemical hydrogen production.

Due to the high IPCE and fast charge transfer with the pyrolysis film, it is natural to apply the pyrolysis hematite film to the dual photoelectrode assembly (31). The pyrolysis film and *p*-GaInP$_2$ electrodes were connected via a multi-meter and immersed in 1M NaOH solution. Unfortunately, zero short circuit photocurrent was obtained at illumination up to approximately 10 suns. To understand this, Figure 7a gives the energy diagram of hematite nanorod film connected with *p*-GaInP$_2$ electrode, assuming the later has the same band edges shift with pH as hematite nanorod. With band gap illumination, the dual photoelectrode system worked and the short circuit photocurrent depended on the photoresponse of hematite nanorod (31). For the system to work, it is critical that the conduction band minimum (CBM) of the n-type semiconductor is higher than the valence band maximum (VBM) of the p-type semiconductor, as shown in Figure 7b, so that the electron could flow in the external circuit. With pyrolysis hematite film, the photocurrent onset shifts *ca.* 0.30 V anodically in the same condition, Figure 4. This shift means that the flat band potential shifts the same way. Due to this shift, the CBM level of the pyrolysis film is below the VBM-(*p*). Therefore, there is no electron flow in the external circuit even in short circuit condition. Owing to the unsatisfied thermodynamic condition, Figure 7b, no short circuit will be obtained. For the device to work, another *p*-type semiconductor with suitable VBM is needed.

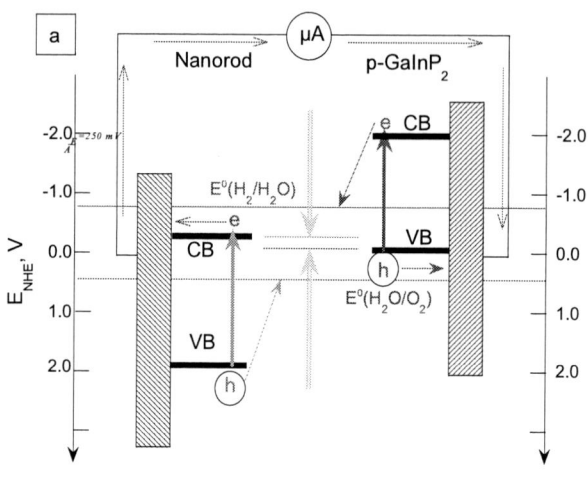

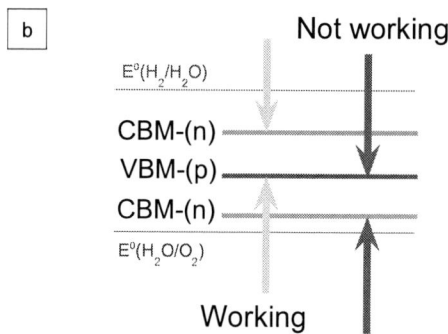

Figure 7. (a) Energy level diagram for the dual-photoelectrode photoelectrochemical water splitting device at pH 14 with a hematite nanorod photoanode and a GaInP$_2$ photocathode. The electron flow direction in the external circuit is shown. (b) Band edge requirements for the dual-photoelectrode assembly to work.

Conclusions

Nanorod and pyrolysis hematite thin films have been investigated. Glancing angle XRD patterns revealed the hematite structure of both, with the pyrolysis film being more compact on the outer-most surface. Optical absorption measurements suggested that both direct and indirect band gap transitions were possible, with the indirect transition dominating. Nanorod film showed a direct optical band gap of 2.14 eV and an indirect optical band gap of 2.04 eV. Pyrolysis film gave a direct optical band gap of 2.15 eV and an indirect band gap of 2.08 eV. Pyrolysis film showed much higher *IPCE* values than those of nanorod film, while nanorod film showed a wider absorption edge. According to these edges, band gaps of 1.94 and 2.00 eV can be estimated for nanorod film and pyrolysis film, respectively.

Photocurrent onsets were –0.44 V for nanorod film and –0.14 V for pyrolysis film, determined by dynamic polarization in 1 M NaOH solution with chopped light. The linear relationship between the photocurrent and the light intensity revealed that a fast charge transfer process is encountered with the pyrolysis film. On the other hand, nanorod film experienced a high charge carriers' recombination that eliminated the external photocurrent. The adoption of pyrolysis film in the dual photoelectrode assembly resulted in zero short circuit photocurrent, which is due to the mis-match of the CBM of the pyrolysis film with the VBM of the p-GaInP$_2$. It is expected that combining the advantages of the two films would result in an efficient electrode for hydrogen production via photoelectrochemical water splitting that have efficient charge transfer and a large surface area.

Acknowledgments

The authors wish to thank Dr. Lionel Vayssieres and Prof. Sten-Eric Lindquist at Uppsala University and Prof. Jan Augustynski at University of Geneva for the hematite films. Dr. Qi Wang at NREL is acknowledged for the assistance in optical measurements. This work was supported by the Fuel Cell Technologies Program of the US Department of Energy.

References

1. A. Fujishima and K. Honda, *Nature,* **238**, 37 (1972).
2. M. Grätzel, *Nature,* **414**, 338 (2001).
3. Y-S. Hu, A. Kleiman-Shwarsctein, A. J. Forman, D. Hazen, J-N Park and E. W. McFarland, *Chem. Mater.,* **20**, 3803 (2008).
4. J. H. Kennedy and K. W. Frese, Jr., *J. Electrochem. Soc.,* **125**, 709 (1978).
5. M. P. Dare-Edwards, J. B. Goodenough, A. Hamnett and P. R. Trevellick, *J. Chem. Soc. Faraday Trans. 1,* **79**, 2027(1983).
6. J. E. Turner, M. Hendewerk, J. Parmeter, D. Neiman, G. A. Somorjai, *J. Electrochem. Soc.,* **131**, 1777 (1984).
7. J. H. Kennedy and M. Anderman, *J. Electrochem. Soc.,* **130**, 848 (1983).
8. V. M. Aroutiounian, V. M. Arakelyan, G. E. Shahnazaryan, H. R. Hovhannisyan, H. Wang and J. A. Turner, *Solar Energy,* **81**, 1369 (2007).
9. E. L. Miller, R. E. Rocheleau, X. M. Deng, *Int. J. Hydrogen Energy,* **28**, 615 (2003).
10. E. L. Miller, D. Paluselli, B. Marsen, R. E. Rocheleau, *Solar Energy Mat. Solar Cells,* **88**, 131 (2005).
11. U. Björkstén, J. Moser and M. Grätzel, *Chem. Mater.,* **6**, 858 (1994).
12. A. Duret and M. Grätzel, *J. Phys. Chem. B,* **109**, 17184 (2005).
13. I. Cesar, A. Kay, J. Matinez and M. Grätzel, *J. Am. Chem. Soc.,* **128**, 4582 (2006).
14. A. Kay, I. Cesar and M. Grätzel, *J. Am. Chem. Soc.,* **128**, 15714 (2006).
15. N. Beermann, L. Vayssieres, S.-E. Lindquist and A. Hagfeldt, *J. Electrochem. Soc.,* **147**, 2456 (2000).
16. T. Lindgren, H. Wang, N. Beermann, L. Vayssieres, A. Hagfeldt, S.-E. Lindquist, *Sol. Energy Mater. Sol. Cells,* **71**, 231 (2002).

17. L. Vayssieres, N. Beermann, S.-E. Lindquist and A. Hagfledt, *Chem. Mater.,* **13**, 233 (2001).
18. X. Wen, S. Wang, Y. Ding, Z. Wang and S. Yang, *J. Phys. Chem. B,* **109**, 215 (2005).
19. R. Schrebler, K. Bello, F. Vera, P. Cury, E. Muñoz, R. del Río, H. G. Meier, R. Córdova and E. A. Dalchiele, *Electrochem. Solid State Lett.,* **9**, C110 (2006).
20. S. U. M. Khan and J. Akikusa, *J. Phys. Chem. B,* **103**, 7184 (1999).
21. C. J. Sartoretti, M. Ulmann, B. D. Alexander, J. Augustynski, A. Weidenkaff, *Chem. Phys. Lett.,* **376**, 194 (2003).
22. C. J. Sartoretti, B. D. Alexander, R. Solarsko, I. A. Rutkowska, J. Augustynski, R. Cerny, *J. Phys. Chem. B,* **109**, 13685 (2005).
23. M. K. Nazeeruddin, A. Kay, I. Rodicio, R. Humphry-Baker, E. Müller, P. Liska, N. Vlachopoulos and M. Grätzel, *J. Am. Chem. Soc.,* **115**, 6382 (1993).
24. A. Hagfeldt and M. Grätzel, *Chem. Rev.,* **95**, 49 (1995).
25. W. W. Gärtner, *Physical Review,* **116**, 84 (1959).
26. V. M. Aroutiounian, V. M. Arakelyan, G. E. Shahnazaryan, E. A. Khachaturyan, H. Wang and J. A. Turner, *Solar Energy,* **80**, 1098 (2006).
27. M. A. Butler, *J. Appl. Phys.,* **48**, 1914 (1977).
28. M. D. Ward, J. R. White, A. J. Bard, *J. Am. Chem. Soc.,* **105**, 27 (1983).
29. H. Wang, T. Lindgren, J. He, A. Hagfeldt and S.-E. Lindquist, *J. Phys. Chem. B,* **104**, 5686 (2000).
30. C. Sanchez, K. D. Sieber and G. A. Somorjai, *J. Electroanal. Chem.,* **252**, 269 (1988).
31. H. Wang, T. Deutsch and J. A. Turner, *J. Electrochem. Soc.,* **155**, F91 (2008).

Photo-Induced Alcohol Electro-Reforming for H₂ Production

A.K.Seferlis[a,b] , S.G.Neophytides[b]

[a] Chemical Engineering Dept., University of Patras
[b] Institute of Chemical Engineering and High Temperature Chemical Processes
(FORTH/ICE-HT), Stadiou str. Platani, GR26504, Rion Patras, Greece
Email address: neoph@iceht.forth.gr

The photoelectrochemical anodic decomposition of alcohols was studied on a porous nanoparticulate TiO_2 film which was illuminated with a 365 nm UV radiation. The open circuit potential (OCP) developed and the produced photocurrent (I_{ph}) was dependent on the nature of the alcohol molecule and on the TiO_2 film thickness. With increasing film thickness the OCP becomes more negative acquiring values varying between -120 mV to -210mV. Correspondingly the photocurrent was enhanced at thicker TiO_2 films and more negative OCP values. This behavior can be attributed to the possibility that characterizes a nanoparticulate porous TiO_2 film, to be penetrated by the electrolyte and depleted from the majority charge carriers (electrons in the conduction band). The optimum film thickness for the given UV radiation depends on the applied positive reversed bias potential and lies, between 3-7µm.

Introduction

The environmental pollution and climate change issues have attracted the attention not only of the scientific community but of the public as well. This was bound to happen once the results of human indifference in environmental pollution and of the exploitation of resources without measure over the past years, all with respect to profit, are becoming more and more evident. The cost humanity is beginning to pay ascends, measuring in human lives but also in billions of dollars. There is also a non measurable cost in the quality of life for every inhabitant of the planet. Inevitably this unfolding problem has shifted the global research to fields such as the clean and renewable energy production and waste processing.

The use of hydrogen as energy carrier is almost certain either in the near future in fuel cells, photo-fuel cells or even in combustion engines, or in the not so near future in nuclear fusion reactors. On the other hand organic waste from industries and civil activities is a problem that must be confronted as soon as possible.

Photoelectrochemical decomposition of organic compounds at semiconductor photoanode has been studied from the 1960s (1). A decade later, it has been reported that a crystalline TiO_2 photoanode can photodecompose water (2) or organic compounds under solar (UV) irradiation. Later on, fine powders were investigated as photocatalysts and titania was proved to be a very efficient photocatalyst for the decomposition of various organic compounds (3-4). The deposition of noble metals, such as Pd, Au and Pt (5–6) has been found to improve performance. Nevertheless the most promising route of hydrogen production processes involves the so-called Photoelectrochemical (PEC) route (7– 9). In this case, the semiconductor photocatalyst is deposited on a conductive substrate electrode (anode) externally connected with a cathode platinum electrode. The photogenerated holes oxidize and decompose the target substance while the electrons move through the external circuit to the counter electrode where they interact with hydrogen ions reducing them and producing hydrogen. This method has the potential to combine a renewable energy resource like sunlight and the reforming of organic substances common to industrial waste in order to provide hydrogen or in other words clean fuel. In the present study the photoelectrocatalytic production of hydrogen was investigated by means of electroreforming of alcohols.

Experimental

All chemicals were purchased from Aldrich and were used without any further purification. Water from a Millipore installation was used in all experimental procedures. Titania films were deposited on Indium – Tin Oxide coated glass slides with 8-12 Ω/sq initial surface resistivity via dip-coating method (10). A solution containing 10.5g Triton X-100, 10.2ml Acetic acid, 57ml pure ethanol and 5.4ml titanium tetraisopropoxide was kept under vigorous stirring. ITO glass slides were initially rinsed with millipore water and ethanol and afterwards calcinated at 550°C in order to remove any impurities from the surface. Subsequently the ITO glass slides were immersed in the solution, so that 5.5cm of the long side of the slide was wetted by the sol, and abruptly withdrawn. The non conductive side of the slide was carefully swept dry and then the slides were calcinated at 550°C for 15 minutes with a heating/cooling rate of 15°C/min. The whole procedure was repeated as required until the desired amount of titania was deposited on the slide. The electrodes were transparent except for samples subject to more than approx. 15 cycles of dip-coating that gradually became white.

The photoreactor (Fig.1) used in all experiments was a Plexiglas cylinder 11.5cm in diameter and 9.5cm in height. A Philips lamp (TL4W/08 blb F8T5) was vertically positioned in the center of the reactor. The lamp emits light mainly in the ultra violet region with a peak at 365nm.The light intensity was measured to be 0.72mW/cm². Two independent Helium gas flow streams were used; one for the main reactor chamber and another for the counter electrode. A delicate adjustment of the flows made possible the sampling of the gas products of the working and counter electrodes. A conventional three electrode system was used with the counter electrode being a platinum wire, while the reference electrode was a stationary hydrogen electrode connected to the reactor via a Luggin capillary. The electrolyte was 0.1M NaOH with various concentrations of ethanol ranging from zero to 3M. SEM images of the various samples were collected by using a LEO SUPRA 35VP electron microscope.

Results

The structural morphology of the samples was examined by means of SEM microscopy. At lower magnifications a cracked – mud morphology was obtained that can be attributed to shrinkage upon drying (11). Fig.2a is a high magnification SEM image of the electrode (top view) revealing a uniform structure consisting of grains with diameter ranging between 20 and 25nm. Fig. 2b shows a cross section of the electrode. The three phases are shown, while it is obvious that there exists no significant difference in the

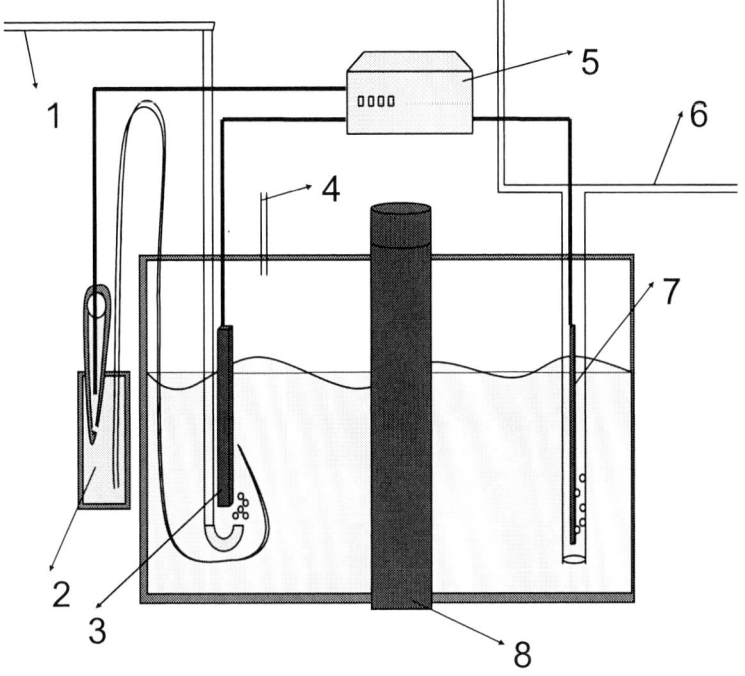

Figure 1. Photelectrochemical reactor setup; 1)Gas Inlet, 2)Reference Electrode, 3)Working Electrode, 4)Gas Outlet, 5)Potensiostat, 6)Counter el. gas inlet /outlet, 7)Counter Electrode, 8)UV lamp

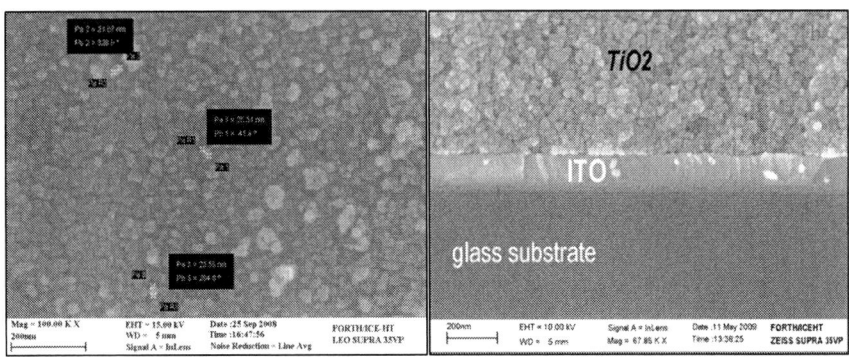

Figure 2. SEM picture of the electrocatalyst top view (left) and crossection (right)

morphology of the samples between successive dip-coating cycles. The XRD spectra, not shown here for brevity, shows that the sample was exclusively anatase with particle size ca. 15nm as estimated by the use of Scherrer equation.

The effect of alcohol presence in the solution on the photocurrent is presented by the I-V plots of Fig.3 for various alcohols with concentration 0.1M. The starting point in each plot is the open circuit potential with respect to H_2 reference electrode (OCP). As the applied potential moves toward positive values the current increases and finally reaches a limiting value. It must be noted that the photocurrent measured corresponds absolutely to the hydrogen produced at the cathode and was verified by gas chromatography measurements. Among all alcohols tested, ethanol gives the most negative OCP and the highest photocurrent, followed by methanol, isopropanol and butanol. The photocurrent at 0V (I_{pho}) corresponds to the spontaneous photo-electro-reforming process (i.e. to the the catalytic rate in the case of the Pt/TiO_2 catalyst), while the saturation photocurrent (I_{phs}), at potentials more positive than 0.6V, to the complete mineralization of the organic compounds (12). The latter can proceed spontaneously by feeding O_2 at the cathode, thus corresponding to the direct alcohol photo fuel cell process. It is obvious that the more negative the OCP is, the higher is the resulting photocurrent. Similar observation has been made for the degradation of the oxalic acid (13).

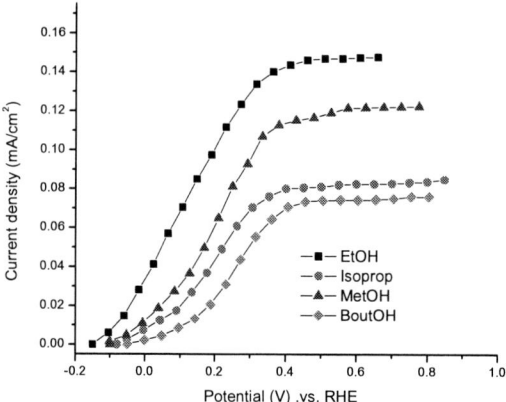

Figure 3. Current density versus potential plots for the first four alcohols in 0.1M concentration, supporting electrolyte was 0.1M NaOH.

Since ethanol appeared to be the most promising the effect of its concentration in the solution was investigated. As presented in fig.4 the photocurrent rises with concentration and that stays true for up to 0.5M. Any further increase results in similar photocurrent values. For reasons of comparison the blank experiment, without ethanol in the solution, is also plotted. Fig.5 depicts the effect of ethanol's concentration on the OCP and on I_{pho} and I_{phs}. All three indicators reach a maximum and then approach a plateau for concentrations above 0.5M of EtOH. The maximum variation of OCP with respect to the non containing EtOH electrolyte is - 450mV, while a tenfold increase in I_{phs} is observed. The shift towards negative OCP and the increase of I_{pho} and I_{phs} with increasing EtOH concentration shows that the semiconductor's built-in voltage, due to band bending

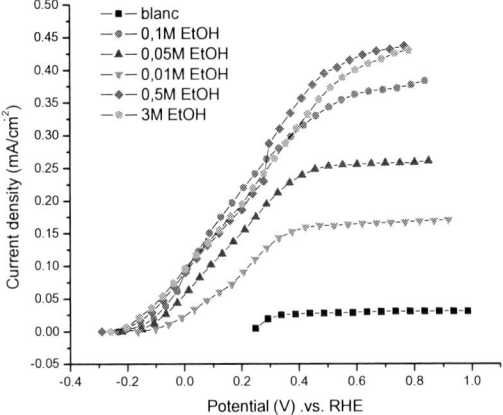

Figure 4. Comparative current density – potential diagram for several ethanol concentrations, supporting electrolyte was 0,1M NaOH.

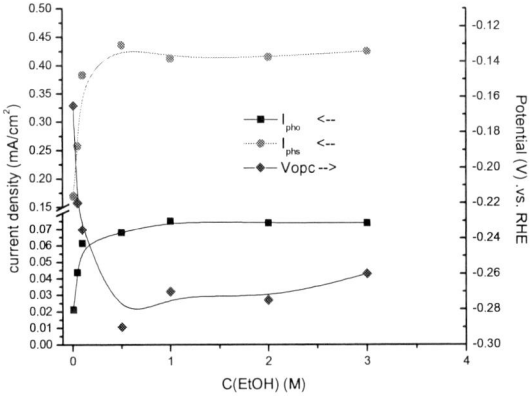

Figure 5. The effect of ethanol's concentration variation in the short-circuit and saturation photocurrents (left axis) and to the open circuit potential (right axis). Supporting electrolyte was 0.1M NaOH.

within the depletion area, may depend on the concentration variation of the redox potential of the EtOH species. Taking into consideration that at the limiting current (I_{phs}) complete charge separation takes place, I_{phs} must be directly dependent by the charge transfer rate of the photogenarated holes to the electrochemical interface which react either directly or indirectly with the organic compound. However the increase of I_{phs} with increasing EtOH concentration can, at the same time, be attributed to the current multiplication effect (14-16) where intermediate radical species can directly inject electrons to the conduction band. The formation of surface states though by the specific adsorption of EtOH or intermediated oxidation species that can facilitate charge transfer cannot be excluded. This is corroborated by the observed constant value of I_{phs} at high EtOH concentrations thus indicating the pinning of the Fermi level of the illuminated TiO_2.

In experiments performed with different amount of titania deposited on the ITO electrodes, at constant ethanol and supporting electrolyte concentration, it has been verified that the increase in the amount of TiO_2 and correspondingly in the film thickness, leads to higher photocurrents (Fig. 6). As it is also shown in Fig. 6 the saturation photocurrent appears at more positive potentials. Certainly the increase of the amount and thickness of the deposited porous nanocrystaline TiO_2 film is expect to result correspondingly in an increase of the electrochemical interface. Thus the increase in I_{phs} is an expected behavior.

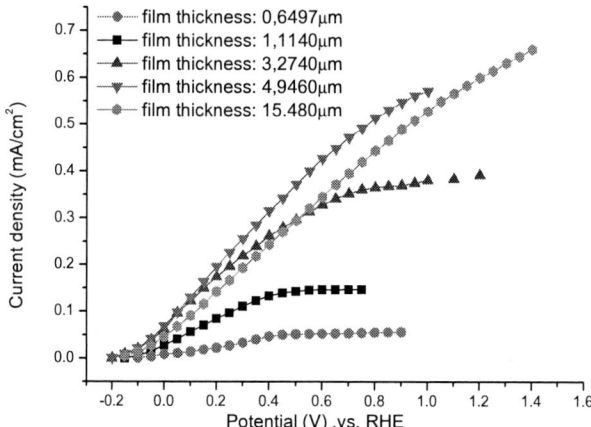

Figure 6. Comparative current – voltage diagram for several electrocatalyst amount, electrolyte was 0,1M NaOH & 0,1M EtOH.

However the most intriguing result is the variation of the OCP with the film thickness shown in Fig. 7. Starting from OCP values around -0.12V it decreases toward negative values with increasing film thickness reaching a shallow minimum at 3-5µm film thickness. This behaviour can be characteristic of nanoporous and nanoparticle thin films most probably related to the diffusion length of electrons within the film matrix (17). At high film thickness the limited penetration absorption length of the incident light can be the reason of the minimum appearing in Fig. 7. Similar behaviour was recently recorded

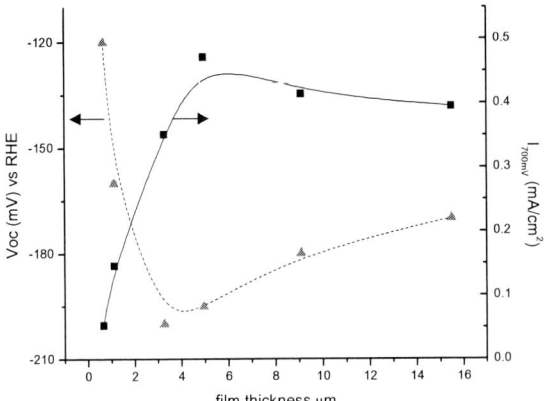

Figure 7. Open Circuit Potential obtained for several electrodes with varying film thickness (left axis), and corresponding current density at 700mV of applied potential (right axis). In all cases electrolyte was 0.1M NaOH & 0.1M Ethanol.

by Krysa et al (13). It must be noted that similar effect is observed for the photocurrent recorded at 0.8V thus implying that similar phenomena affect both OCP and photocurrent.

Discussion

Though significant theoretical work was devoted in the past in order to get a better understanding of the behavior of the porous nanocrystalline semiconductor electrochemical interface our knowledge is still limited (18). Hagfeldt et al (19) proposed the existence of an optimum porous nanocrystalline film thickness of approximately 4 μm which corresponds to a maximum overall quantum efficiency. Theoretical models for the action spectrum and the current-voltage characteristics of microporous (colloidal) semiconductor films in photoelectrochemical cells have been derived by Lindquist et al (17). Their model was based on a 3-dimensional isotropic microporous network consisting of nano-sized low-doped TiO_2 semiconducting nanoparticles. Therefore the electrolyte is able to penetrate the film all the way to the back contact, and every nanoparticle being in contact with the electrolyte will be totally depleted by the major charge carriers (20). This will totally deplete all charge carriers throughout the bulk of the semiconductor's film. Thus, no band bending is expected to occur within both the nanoparticles and the bulk of the TiO_2 film. In this respect the charge transfer will be limited by the diffusivity of electrons toward the back contact. They estimated an electron diffusion length of 0.8μm based on the assumptions that the major charge carrier (electrons) transport in the semiconductor occurs via diffusion, and that the diffusion length is constant. However the electron diffusion coefficient will certainly depend on the applied potential and is expected to vary along the film thickness.

Based on the above considerations it must be noted that for the case of the thinner samples (0.65 and 1.11μm) the limiting factor that determines the charge transfer rate can be the rate of formation and transfer of the light induced generated holes (minority charge carriers) to the electrochemical interface. This is corroborated by the appearance of the saturation photocurrent at the same potential, which is enhanced proportionally by the increase of the film thickness and thus of the amount of deposited TiO_2.

As shown in Fig. 6 the I-V curves for the samples with film thicknesses 3.27μm and 4.95μm essentially coincide within the potential range from OCP to 0.2V, while the performance of the thicker electrode is enhanced with respect to the thinner one depending on the applied reversed bias potential. This can be attributed to the fact that at more positive potentials the exploitation of the electrochemically active surface area is higher. The latter can be rationalized by the dependence of electron diffusivity on the applied potential. At low reversed bias potential, the charge transfer electrochemical processes are limited by the diffusivity of electrons across the film which is fixed. In this respect only up to a certain fixed portion of the electrochemically active surface area will be exploited no matter which is the TiO_2 film thickness. Thus electrochemical active sites close to the outer film surface being far away from the layers closer to the conducting substrate, will not participate in the photocurrent. On the other hand at more positive reversed biased potentials the stronger effect of the applied electric field on the major charge carrier transfer (diffusivity) across the TiO_2 film will allow the outer electrochemically active surface layers to participate in the charge transfer electrochemical processes.

At even larger film thickness the photocurrent generally drops, although as it is shown in Fig. 6, at higher positive reversed bias potentials (1V) the photocurrent does not approach yet its saturation value. This behavior can be rationalized by considering the limited penetration depth of light in the nanoporous polycrystalline film (17). As the sample is illuminated from the side of the surface that is interfaced with the electrolyte it is reasonable to assume that the part of the film positioned closer to the substrate current collector will not be light driven, while the photocurrent from the outermost layers is limited by the diffusion length of the majority carriers (electrons).

The variation of the OCP with film thickness must be closely related to the semiconductor/substrate contact and the high degree of the depletion of the major charge carriers (no band bending) in the bulk of the nanoparticles and consequently across the film. The highest negative value of the OCP of n-type illuminated semiconductor is obtained at the band edge at the surface of the semiconductor. The voltage variation along the depletion area will determine the OCP value measured with respect to a reversible reference electrode. If the size of a nanoparticle varies within the dimensions of the space charge region (depletion area) the voltage in the bulk of the nanoparticle will become more negative with decreasing particle dimensions. At particle size as large as 30nm the variation of the voltage within the particle is minimal (21) thus approaching the highest negative value, which is equal to the voltage value at the particle's electrolyte interface. Taking into consideration that the porous TiO_2 film is composed of nanoparticles wetted by the electrolyte it can be easily realized that the bulk of the film all the way to the substrate current collector is within the dimensions of the space charged region and the voltage in the bulk is approaching the voltage developed at the interfacial region. However due to the back contact of the film with the current collector a charge equilibration will be settled between the illuminated semiconductor film and the substrate. In particular electron flow to the depleted charge area will result in the shift of its voltage to more positive values with corresponding shift of the Fermi level. This phenomenon is expected to be more intense for the thin films, where the Fermi level shift will be more significant as compared to thicker films. As depicted in Fig. 7, with increasing film thickness up to 3 - 5μm, the OCP becomes more negative by 80mV which after the minimum approaches to a constant value at -170mV. The minimum observed in Fig. 7 must be related to the absorption depth of the incident radiation. The reciprocal absorption length (α) of the incident radiation can be estimated from the approximation given in (17) shown by eq. 1:

$$\ln(\alpha \ \mu m^{-1}) = 29 - 85 \ \lambda \ \mu m \qquad [1]$$

Based on eq. 1 the absorption length at 365nm is ca. 7μm. At film layers thicker than the absorption length an interfacial area between the illuminated and dark areas of the TiO_2 film will exist which will not be affected by the charge equilibration at the TiO_2/substrate contact. Shiga et al (22) came to the conclusion that the variation in the work function of the substrate does not affect the performance of the porous TiO_2 nanoparticulate film. They proposed that at the TiO_2/electrolyte/substrate interface a Fermi level pinning can be considered which is due to surface states that are developed there, because of the penetration of the electrolyte through the semiconductor thin film being in contact with the conducting substrate as well as the semiconductor thin film.

Conclusions

The introduction into the solution of ethanol concentrations as low as 0.5M, results in one order of magnitude larger hydrogen production rates. Increase of the photoanode film thickness greater than 3-5μm leads to more efficient photoresponse and increased photocurrent.

This method has certain advantages such as the low cost of materials used, the fully controlled reaction rate from the variation of the applied potential that results in the production of 100% hydrogen in the gas stream of the cathode, the 100% reforming of the organic compounds used and that the CO_2 produced is enclaved in the solution and not released in the atmosphere. It has a great potential for application in efficient photo-fuel cell devices for simultaneous waste water treatment and hydrogen production.

Acknowledgments

This work is part of the 03ED375 research project, implemented within the framework of the 'Reinforcement Programme of Human Research Manpower' (PENED) and co-financed by National and Community Funds (20% from the Greek Ministry of Development – GSRT and 80% from E.U. European Social Fund.

References

1. H. Gerischer, *Z. Phys. Chem.* **26**, 325(1960)
2. K. Honda, A. Fujishima, *Nature* **238**, 37(1972)

3. A. Fujishima, K. Hashimoto, T.Watanabe in *TiO2 Photocatalysis—Fundamental and Applications*, BKC Inc., Tokyo, (1999.)
4. K. Kalyanasundaram, M. Graetzel (Eds.) in *Photosensitization and Photocatalysis Using Inorganic and Organometallic Compounds*, Kluwer Academic Publishers, Dordrecht, (1993).
5. D.Kondarides et al, *Catal. Lett.* **122**, 26 (2008)
6. M. Matsuoka et al. *Catal Today* **122**, 51 (2007)
7. M. Kaneko et al, *Electrochem Commun* **8**, 336 (2006)
8. T. Bak et al, *Int J Hydrogen Energy* **27**, 991 (2002)
9. O. Varghese ey al, *Sol Energy Mater Sol Cells* **92**, 374 (2008)
10. N. Strataki et al. *Appl.Cat.B*, **77**, 184 (2007)
11. J. Georgieva et al. *Electroch. Acta* **51**, 2076 (2006)
12. H. Zhao et al. *Journal of Catalysis* **250**, 102 (2007)
13. J. Krysa et al *Elecroch.Acta* **50**, 5255 (2005)
14. N. Hykaway et al., *J. Phys. Chem.*, **90 (25)**, 6663 (1986)
15. T.L.Villarreal et al, *J. Phys. Chem. B*, **108 (39)**, 15172 (2004)
16. D. Jiang et al, *Environ. Sci. Technol.*,**41 (1)**, 303-308 (2007)
17. S. Sodergren et al *J. Phys. Chem.*, **98**, 5552 (1994)
18. L. M. Peter and D. Vanmaekelbergh in *Advances in Electrochemical Science and Engineering*, Edited by Richard C. Alkire and Dieter M. Kolb, WILEY-VCH Verlag GmbH, 1999
19. A. Hagfeldt et al., *Solar Energy Materials and Solar Cells*, **27,** 293 (1992)
20. W.J. Albery et al., *J. Electrochem. Soc.,* **131,** 315 (1984)
21. B. O'Regan, J. *Phys. Chem.,* **94,** 8726 (1996)
22. A. Shiga et al., *J. Phys. Chem. B* **102,** 6049 (1998)

ECS Transactions, 25 (42) 73-82 (2010)
10.1149/1.3416203 ©The Electrochemical Society

Photoelectrolysis of Water in Tj-a-Si Solar Cell Biased CM-n-TiO₂ | | Pt and in monolithic Self-Driven n-TiO₂ – Mn₂O₃ coated Tj-a-Si | | Pt Photoelectrochemical Cells

M. Frites[a], W. B. Ingler[b] and S. U. M. Khan[a]*

a Department of Chemistry and Biochemistry, Duquesne University, Pittsburgh, PA
15282
b Department of Physics, University of Toledo, Toledo, OH, 43606
* email: khan@duq.edu

Carbon modified (CM)-n-TiO₂ thin film was synthesized and optimized by flame oxidation method at high temperature of 825°C, and 16 min heating time. The natural gas and oxygen flow rates of 3.32 L/min and 2.23 L/min were found optimum to incorporate the appropriate amounts of carbon concentration in the bulk and on the surface of the CM-n-TiO₂ thin films. TJ-a-Si solar cell biased CM-n-TiO₂ | 2.5 M KOH | Pt PEC generated hydrogen gas of 25.0 L m⁻² h⁻¹ and a photoconversion efficiency of 5.45 % under the solar simulated light illumination of 1 sun. Importantly, the Tj-a-Si solar cell biased dark water electrolysis cell of Pt | 2.5 M KOH | Pt in did not show any measurable H₂ gas production. The monolithic Self-Driven n-TiO₂ –Mn₂O₃ coated Tj-a-Si | 2.5 M KOH | Pt PEC generated hydrogen gas of 15.63 L m⁻² h⁻¹ and a photoconversion efficiency of 4.16 %. This PEC was found to last more than 5 hours (compared to maximum 5 min life time for uncoated Tj-a-Si | 2.5 M KOH | Pt PEC). This limited stability for 5 hours was attributed to presence of some pin-holes on n-TiO₂–Mn₂O₃ coated Tj-a-Si solar cell surface.

Introduction

The drive to develop clean, renewable sources of energy is in high demand with the world's growing dependence on fossil fuel as the primary energy source. Fossil fuel when burned, emit harmful by-products, greenhouse gas such as CO₂ which contributes to global warming, sulfur gases and mercury released from coal fired power plant are directly associated with acid rain, and acid mine drainage. The photoelectrochemical water splitting on the semiconductor electrode surface converting solar energy to hydrogen fuel offers an alternative energy solution which is clean and renewable (1). Progress in R&D on photo-electrochemical hydrogen generation is based on the development of new photo-active materials and photoelectrode fabrication. Oxide based semiconductors appear to be the most promising materials for water splitting using solar energy due to their long-term chemical stability, non toxicity, and relatively low cost processing technology. Moreover, their properties can be chemically modified by incorporating impurities such as H, N, S, C, and transition metals in their crystalline structures (2-9). Since the first publication on amorphous silicon (a-Si) relevant to the solar cell in the late 1960s (10), it has been studied as a promising material not only for electricity generation but also for water splitting to hydrogen and oxygen gases. Amorphous Silicon (a-Si) is cheaper than its crystalline form and the stacked triple-junction thin-film (nip/nip/nip) amorphous silicon (Tj-a-Si) generates higher photovoltage and photocurrent density. The a-Si cell also has a lower light induced degradation; however, it lacks stability in aqueous solutions. Different materials and

73

techniques have been employed for Si surface passivation in electrolyte solutions to improve its stability during oxygen evolution (11-13).

Carbon modified n-TiO$_2$ (CM-n-TiO$_2$) or regular n-TiO$_2$ thin films can be deposited or transferred to the surface of Tj-a-Si, by vacuum deposition or electrochemical deposition to fabricate a state of the art stable Tj-a-Si for efficient photochemical splitting of water to H$_2$ and O$_2$. The cost of solar to hydrogen production can be cut to a minimum if one uses a monolithic system that eliminates costs associated with separate construction and interconnection of solar cells and electrochemical water electrolyzers. The concept is to use Tj-a-Si as photoanode or photocathode where the oxygen evolution reaction (OER) or hydrogen evolution reaction (HER) occurs on its surface depending on n-type or p-type topmost layer. The use of such system is hampered by the photo-degradation and therefore unsuitable for long term use (12). Recently, thinner films became accessible with the use of dip coating, spin coating, CVD (chemical vapor deposition) and electro-deposition techniques (9, 14-17). To protect Tj-a-Si one use these methods to deposit a transparent conducting corrosion resistant (TCCR) n-TiO$_2$, CM-n-TiO$_2$, n-Fe$_2$O$_3$ or CM-n-Fe$_2$O$_3$ layers on the top of Tj-a-Si at low temperature (< 250 °C). Note that it is critical to have high attachment, stability and pin-hole free surface. The goal is to deposit TCCR layer on Tj-a-Si and investigate the stability and electric contact between the TCCR thin film and the Tj-a-Si surface in the electrolyte solution.

In this paper we report the Photoelectrolysis of water to hydrogen and oxygen in Tj-a-Si Solar Cell Biased CM-n-TiO$_2$ | 2.5 M KOH | Pt and in monolithic Self-Driven n-TiO$_2$ – Mn$_2$O$_3$ coated Tj-a-Si | 2.5 M KOH | Pt Photoelectrochemical Cells (PECs). We also studied the stability of titanium oxide coated Tj-a-Si solar cell during direct water splitting reaction in a monolithic self-driven Tj-a-Si | 2.5 M KOH | Pt photoelectrochemical cells (PEC).

Experimental Details

Synthesis of CM-n-TiO$_2$ thin films by flame oxidation:

Ti metal sheets of 0.25 mm thick (Alfa Co.) were cut to an area of ~ 1.0 cm^2. Ti metal samples were cleaned in a sonicator for three-fifteen minute intervals with: (i) Acetone; (ii) acetone: double de-ionized water (1:1); (iii) double de-ionized water. The cleaned Ti metal sheets were flame oxidized at 825°C and 16 min heating time using a custom designed large area flame (Knight, Model RN 3.5xaWC) under controlled natural gas and oxygen gas flows. A digital thermocouple (Thermolyne Co. Model PM-20700) was used to measure the flame temperature near the metal sheet. The temperature of the flame was critical and maintained by controlling the flow rates of oxygen and natural gas. The oxygen gas flow rate was measured by using an FL-1807 flow meter (Omega Engineering) and natural gas flow rate by an FL-1806 flow meter (Omega engineering). The natural gas and oxygen flow rates of 3.32 L/min and 2.23 L/min were found optimum to incorporate the appropriate amounts of carbon concentration in the bulk and on the surface of the CM-n-TiO$_2$ thin films

The photoelectrodes were prepared by removing the oxide layers in the top portion on both back and front sides of the Ti metal sheet using a file to make electrical contact. The back side and part of the front side were covered with a nonconductive epoxy adhesive. The uncovered area of the photoelectrodes (~ 0.2 cm^2) was calculated using a computer program "Image J". Electrical connection was made using a copper clip to the exposed surface of Ti metal in the upper region of the photoelectrode. A single compartment electrochemical cell containing 2.5 M KOH solution was used to carry out the water splitting reaction under illumination of light from an Oriel Solar Simulator. The rate of water splitting reaction was obtained in terms of photocurrent density, j_p (mA/cm^2) calculated from the volume of hydrogen gas collected over the electrolyte solution of 2.5 M KOH.

Oxygen gas collection at CM-n-TiO$_2$ as photoanode and H$_2$ gas collection at Pt cathode biased by Tj- a-Si solar cell in CM-n-TiO$_2$ | 2.5 M KOH | Pt PEC:

A single compartment electrochemical cell with CM-n-TiO$_2$ as photoanode for oxygen evolution and Pt wire as dark cathode for hydrogen evolution was used. The required minimal bias was provided by the Tj- a-Si solar cell under light illumination of 100 mW/cm^2 from a solar simulator (see Figure 1). Both Tj-a-si (placed above the solution) and the photoanode, CM-n-TiO$_2$ were illuminated with the same intensity of light from the solar simulator. The electrical connection was made so that the positive surface of the solar cell is connected to the anode and the negative back side of the solar cell which consists of stainless steel is connected to the Pt wire for HER. Hydrogen gas was collected by inverting the electrolyte filled graduated test tube on the top of Pt wire as shown in Figure.1. Hydrogen gas evolution was monitored by displacement of the electrolyte. The same method was used to monitor the oxygen evolution on the photoanode and verified the volume of H$_2$ and O$_2$, ratio, V_{H2} / V_{O2} to be 2 as expected

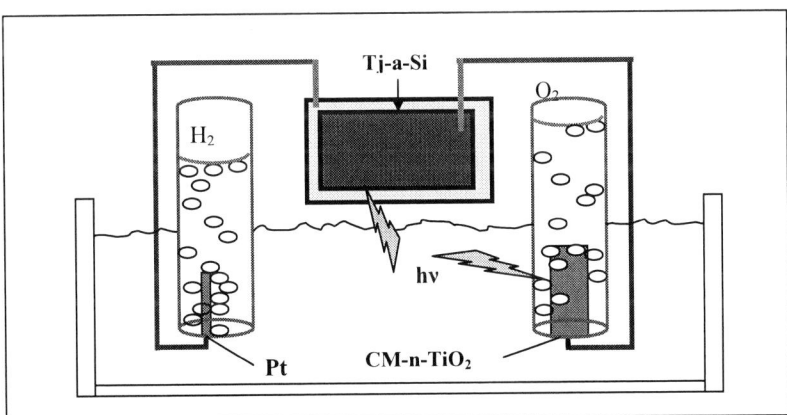

Figure 1. A Schematic view of the photoelectrolysis using an electrochemical cell system with CM-n-TiO$_2$ as photoanode and Pt metal as dark cathode biased by Tj-a-Si solar cell. Deposition of n-TiO$_2$ and Mn$_2$O$_3$ thin films on Tj-a-Si substrate to fabricate n-TiO$_2$-Mn$_2$O$_3$ coated Tj-a-Si solar cell:

A new and advanced method to deposit a homogenous and durable thin film of n-TiO$_2$ on the surface of Tj-a-Si surface was used in which an ultrafine colloidal solution of anatase n-TiO$_2$ (Degussa p 25) instead of the suspension was utilized. The colloidal solution consists of 80 mg of anatase n-TiO$_2$ powder in 10 mL ethanol, in presence of a structure directing surfactant, acetylacetone (9) or polyethylene glycol (PEG) (18). A binder Triton X100 (5 micro liters in 10 mL ethanol) was added to the suspension to obtain homogeneous films with good mechanical and electrical properties so that no cracking of the film occurs as it dries. The mixture was first homogenized in a high speed centrifuge (Avanti J-20XP centrifuge) at 3,000 rpm for 15 min giving an almost clear colloidal solution at the top and a white precipitate (anatase powder) at the bottom of the container. Multiple drops from the top of the ultrafine colloidal solution were spread on the surface of Tj-a-Si using a glass rod then spin coated for 1 min then dried at room temperature. The resulting films were sintered in an electrical oven at 200°C for 1 hour to obtain a uniform solvent free n-TiO$_2$ thin film. The deposition bath for Mn$_2$O$_3$ was prepared from a solution containing 0.03 M MnCI$_2$, 0.25M NH$_4$Cl, and 1.4 M NH$_4$OH aqueous solution having total volume of 50 mL. The solution was stirred continuously and maintained at room temperature. After a few minutes, a light brown precipitate of Mn(OH)$_2$ started forming. The corrosive ammonium chloride was removed from the solution by multiple centrifugation, the resulting Mn(OH)$_2$ brown precipitate was first washed with deionized water (18 Ω) and then homogenized in a sonicator with deionized water with same volume (50 mL) as that of the original deposition bath. The (Tj-a-Si/n-TiO$_2$) sample was then immersed in the deposition bath and left for 15 min. The sample was rinsed with double deionized water and after drying at room temperature, it was transferred to a glass funnel that was evacuated to 2×10^{-5} torr. The sample was then heated at 220 °C for 15 min. The heating in vacuum resulted in loss of water, thereby leaving a thin film of manganese oxide (Mn$_2$O$_3$) according to the reactions 1 and 2 (19):

$$MnCl_2 \text{ (aq)} + 2 NH_4OH \text{ (aq)} \longrightarrow Mn(OH)_2 \text{ (s)} + 2NH_4Cl \text{ (aq)} \quad [1]$$

During heating at 220 ^{0}C and under 2×10^{-5} torr:

$$2 Mn(OH)_2 \text{ (s)} \longrightarrow Mn_2O_3 \text{ (s)} + H_2 \text{ (g)} + H_2O \text{ (g)} \quad [2]$$

Formation of Mn$_2$O$_3$ oxide was previously confirmed earlier by XPS analysis (12)

Hydrogen gas collection at Pt cathode in n-TiO$_2$-Mn$_2$O$_3$ coated Tj-a-Si | 2.5 M KOH | Pt PEC:

A single compartment electrochemical cell with n-TiO$_2$-Mn$_2$O$_3$ coated Tj-a-Si as photoanode for OER and Pt wire as dark cathode for HER was used. The required minimal bias was provided by the n-TiO$_2$-Mn$_2$O$_3$ coated Tj-a-Si solar cell itself under light illumination of 100 mW/cm^2 from a solar simulator (see Figure 5). The electrical connection was made so that the negative back side of the solar cell which consists of stainless steel is connected to the Pt wire for hydrogen evolution. Hydrogen gas was collected by inverting the electrolyte filled graduated test tube on the top of Pt wire as shown in Figure. 2. The hydrogen gas was collected by displacement of the electrolyte. Oxygen gas evolution occurred on the positive surface of the n-TiO$_2$-Mn$_2$O$_3$ coated Tj-a-Si solar cell.

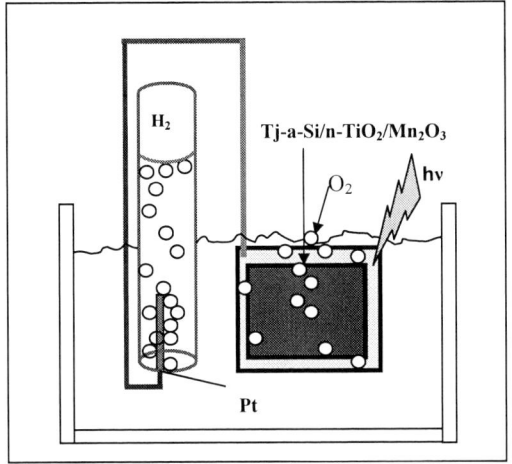

Figure. 2 Schematic view of the monolithic Self-Driven n-TiO$_2$–Mn$_2$O$_3$ coated Tj-a-Si | 2.5 M KOH | Pt PEC Tj-a-Si solar cell system for self-driven generation of hydrogen and oxygen gases by water splitting.

Result and Discussion

All Solar Energy Driven hydrogen production in CM-n-TiO$_2$ | 2.5 M KOH | Pt PEC biased by Tj-a-Si solar cell

Volumes of H$_2$ gas collected

Figure 1 shows the experimental set up for all solar to hydrogen production by photoelectrolysis using CM-n-TiO$_2$ as photoanode and Pt metal as the dark cathode in a single compartment CM-n-TiO$_2$ | 2.5 M KOH | Pt PEC biased by Tj-a-Si solar cell placed outside the electrolyte solution. A 2.5 M KOH solution was used as the electrolyte in this single PEC. CM-n-TiO$_2$ photoanode and Tj-a-Si solar cell were placed at a distance of 7 inch from a solar simulator with global AM 1.5 filter having light intensity of 100 mW cm^{-2} (1 sun).

The collected H$_2$ gas volume was found to be 0.5 mL in 60 min at 0.20 cm^2 area of photoelectrode. This corresponds to a volume of 0.69544 μL H$_2$ cm^{-2} sec^{-1} which corresponds to $\sim$ 25.0 L m^{-2} h^{-1}. This volume of hydrogen gas collected by the displacement of electrolyte solution (taking into account the partial pressure of water vapor as 23.56 mm Hg at 25 °C) was converted to moles of H$_2$, n_{H2} = 0.027507 μ moles H$_2$ cm^{-2} sec^{-1} by using the ideal gas law, n_{H2} = PV_{H2}/RT. The volume of H$_2$ gas collected as a function of time is given in Figure 3. Note that the volume of O$_2$ gas collected was found to be exactly half of that of H$_2$ gas as expected.

Fig. 3 is not linear because the initial rate is high with high slope then it slows down indicating lower rate.

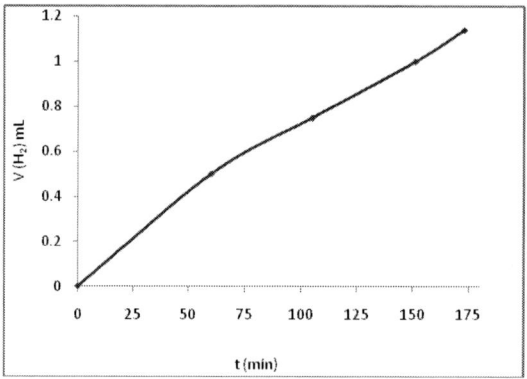

Figure 3.   Volume of hydrogen gas collected versus time

Photocurrent density from moles of H_2 gas collected:

These moles of hydrogen gas (n_{H2}) can be converted to photocurrent density, j_p using the expression,

$j_p = 2n_{H2}F$
$\quad = 2 \times 0.027507 \ \mu \ mol \ H_2 \ cm^{-2} \ sec^{-1} \times 0.096487 \ C \ (\mu \ mol)^{-1} = \ 5.3 \ mA \ cm^{-2}$ [3]

where F is the Faraday constant which is 96,487 coulomb mol^{-1} or 0.096487 C $(\mu \ mol)^{-1}$. The plot of photocurrent density calculated from volume of hydrogen gas collected as a function of time is shown in Fig. 4 where it reaches to a peak value 5.17 mA cm^{-2} in 60 min. then falls to a steady state regime. This behavior can be attributed to the steady state coverage of the semiconductor surface by tiny oxygen gas bobbles after 60 min and can be minimized by occasional chopping of the light.

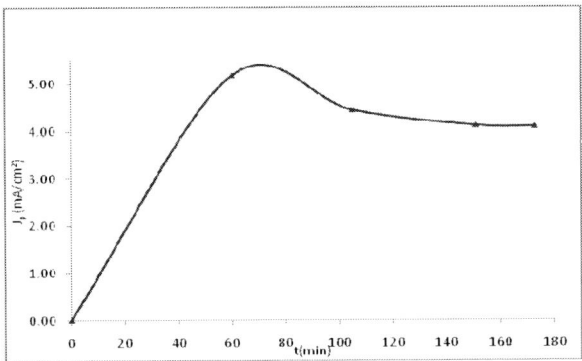

Figure 4. A Plot of photocurrent density, j_p values from volume of Hydrogen gas collected in CM-n-TiO$_2$/Pt PEC during photosplitting of water as a function of time. The power density of 19.47 mW cm^{-2} was provided by a Tj-a-Si solar cell.

Photocurrent density vs. Photovoltage for Tj-a-Si solar cell:

Figure 5 shows the photocurrent density-photovoltage plot for Tj-a-Si solar cell. A photovoltage 1.7 V was found at open circuit condition under illumination of 100 mW cm^{-2}. Under close circuit condition a maximum photocurrent of 19.5 mA cm^{-2} and the corresponding photovoltage of 1.0 volt were observed as shown in Figure 5. The maximum power density of Tj-a-Si solar cell, $P_{Solar\ cell}$ was found to be 19.47 mW/cm^2 as shown in Figure 5.

Photoconversion Efficiency :

For the Tj-a-Si solar cell biased CM-n-TiO$_2$ | 2.5 M KOH | Pt PEC, the photoconversion efficiency can be expressed as,

Photoconversion Efficiency (%) = [(j_p x E^0_{rev})/ (P_{light} + $P_{solar\ cell}$)] x 100
  = [5.3 mA cm^{-2} x1.23 V)/ (100 +19.47) mW cm^{-2})] x 100
  = 5.45   [4]

where I_{light} = 100 mW/cm^2 is the power density of white light from the solar simulator. Using the observed photocurrent density, j_p = 5.17 mA cm^{-2} for the Tj-a-Si biased CM-n-TiO$_2$ | 2.5 M KOH | Pt PEC (obtained from the collected volume of hydrogen gas using Eq. 3), the standard reversible potential for water splitting reaction, E_{rev} = 1.23 V and maximum power supplied by solar cell, $P_{solar\ cell}$ = 19.47 mW cm^{-2} in Eq. 4, the photoconversion efficiency was found to be 5.45 %. The Photoconversion efficiencies can be calculated also from the heat of combustion, ΔH^0 of H$_2$ gas collected per unit time can be calculated as follows:

Photoconversion Efficiency (%) = [(ΔH^0 x n_{H2})/ (P_{light} + $P_{solar\ cell}$] x 100

= [(0.24182 j µmol-1 x 0.027507 µmol cm^{-2} s^{-1})/(100 + 19.47)mW cm^{-2}] x100

= 5.57%   [5]

ΔH^0 = - 241.82 kJ mol^{-1} is the heat of combustion reaction of H_2 gas,

$$H_2\ (g) + \tfrac{1}{2}O_2\ (g) \rightarrow H_2O\ (g) \qquad \Delta H^0 = - 241.82\ kJ \qquad [6]$$

Note that the difference in values of photoconversion efficiencies in Eqs. 4 and 5 arises for the reason that 1.23 volt used in eq. 4 is obtained from the free energy of reaction needed to form H_2 (g) and O_2(g) from liquid water, H_2O (l). However, ΔH^0 value used in Eq. 5 is the heat of combustion reaction for H_2 (g) to form gaseous water, H_2O (g).

It is important to note that the similar experimental set up as shown in figure.2 was used for conventional electrolysis using Pt based dark anode and dark cathode biased by Tj-a-Si solar cell but no measureable H_2 or O_2 gas evolution was observed.

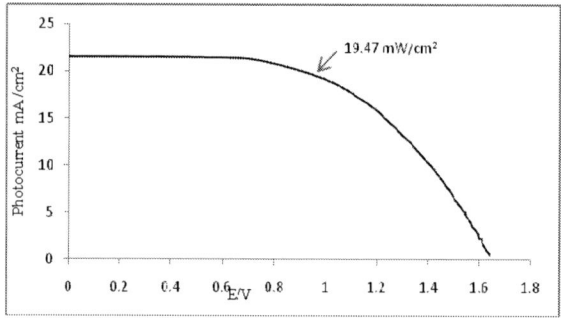

Figure 5. Plot of photocurrent density vs. photovoltage for Tj-a-Si solar cell to determine its aximum power density.

Direct Photosplitting of Water to H_2 and O_2 gases at self-driven monolithic n-TiO$_2$ – Mn$_2$O$_3$ Coated Tj-a-Si | 2.5 M KOH | Pt PEC

The resulting n-TiO$_2$ – Mn$_2$O$_3$ coated Tj-a-Si solar cell was tested in a monolithic system for water electrolysis in 2.5 M KOH solution as shown in the schematic PEC in Fig. 2. The back stainless steel was electrically connected to a Pt electrode and then covered with a non- conductive epoxy. The monolithic system was under illumination intensity of P_o = 100 mW cm^{-2} from a Solar simulator with a global AM 1.5 filter.

The photoconversion efficiency of light-to-hydrogen production was calculated from the collecting the H_2 gas over the electrolyte solution in an inverted graduated glass tube. A photocurrent density of 3.31 mA/cm^2 was found from the volume (0.5 mL) of H_2 gas collected in 32 min at 0.6 cm^2 area of Tj-a-Si solar cell (= 1.5625 mL cm^{-2} h^{-1}) and this generated a photoconversion efficiency of 4.0 % for H_2 production obtained from the relation,

$$\text{Photoconversion efficiency (\%)} = \frac{\Delta H^\circ \times n_{H_2}}{P_{light}} \times 100 \qquad [7]$$

$$= [0.241820 \text{ j } \mu mol^{-1} \times 0.0172 \ \mu mol \ cm^{-2} \ s^{-1} \times 100]/0.1 \text{ W } cm^{-2}$$
$$= 4.16 \ \%`$$

Stability of n-TiO$_2$–Mn$_2$O$_3$ coated Tj-a-Si Solar Cell

Figure 7 shows the Stability of n-TiO$_2$–Mn$_2$O$_3$ coated Tj-a-Si Solar Cell during hydrogen and oxygen generation in the n-TiO$_2$–Mn$_2$O$_3$ Coated Tj-a-Si │ 2.5 M KOH │ Pt PEC under illumination of 1 sun from an AM 1.5 solar simulator. δV in this Figure 7 represents the amount of the volume of H_2 gas collected per cm^2 of photoelectrode (n-TiO$_2$–Mn$_2$O$_3$ coated Tj-a-Si Solar Cell) in every 3.5 min. This amount remained constant up to 5 hours and started to decline in next half an hour to zero volume. This indicates that the n-TiO$_2$–Mn$_2$O$_3$ Coated Tj-a-Si solar cell was stable for 5 hours which showed 60 times more stable compared to uncoated Tj-a Si solar cell that lasted only 5 minutes in 2.5 M KOH during water splitting reaction. Pin holes in TiO$_2$-Mn$_2$O$_3$ coat on Tj-a-Si surface may be responsible for the degradation of the sample after 5 hours. Note that only n-TiO$_2$ coated Tj-aSi solar cell survived maximum 4 hours.

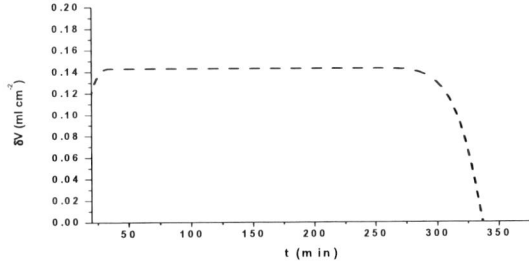

Figure 6. A plot of volume of hydrogen gas collected in every 3.5 min, δV (in mL cm^{-2}) per unit surface area of n-TiO$_2$-Mn$_2$O$_3$ coated Tj-a-Si solar cell Vs time, in n-TiO$_2$-Mn$_2$O$_3$ coated Tj-a-Si │ 2.5 M │ Pt PEC.

Conclusion

In Conclusion, it should be pointed out that flame synthesized carbon modified n-TiO$_2$ (CM-n-TiO$_2$) could efficiently photosplit water to hydrogen and oxygen (> 5.0 %) when biased by a triple junction amorphous silicon (Tj-a-Si) solar cell. When Tj-a-Si is coated by a corrosion resistant n-TiO$_2$–Mn$_2$O$_3$ it can directly photosplit water efficiently (4.0 %) to H_2 and O_2 gases without the need of external bias potential. The transparent n-TiO$_2$–Mn$_2$O$_3$ layer could protect jJ-a-Si for 5 hours in 2.4 M KOH solution under illumination of 1 sun from a solar simulator compared to 5 min for bare Tj-a-Si. Long term stability can be achieved if completely pin hole free transparent corrosion resistant layer could deposited on Tj-a-Si solar cell.

Acknowledgement

We gratefully acknowledge the financial support of US Air force to Duquesne University via a subcontract from University of Toledo.

References

1. T. N. Vezirouglou, Dawn to the hydrogen age. Int. J. Hydrogen Energy. **23**, 1077 (1998)
2. D. Dvoranova, V. Brezova, V. Mazur, M. Malati, Appl, M. A. Catal. B: Env., **37**, 91 (2002)
3. C. F. Chi, Y. L. Lee, H. S. Weng, Nanotechnology, **19**, 125704 (2008)
4. T. L. Thompson, J. T. Yates, J. Chem. Rev., **106**, 4428 (2006)
5. H. Wang, J. P. Lewis, J. Phys, Condens. Matter, **18**, 412 (2006).
6. D. C. Cronomeyer, Phys . Rev., **13**, 1222 (1959).
7. S. U. M. Khan, M. Al-Shahry, and W. Ingler Jr Science., **297**, 2243 (2002)
8. M. Frites, Y. A. Shaban, S. U. M. Khan, in Dielectric and semiconductor materials, devices, and processing, p. 85, The Electrochemical Society Proceedings Series, Pennington, NJ (2008).
9. H. Kisch, W. Macyk, Chemiphyschem, **3**, 399 (2002).
10. A. Goetzberger, V.U. Hoffmann, Photovoltaic Solar Energy Generation. Springer, (2005).
11. G. H. Lin, M. Kapur, R. C. Kainthla, and J. O'M. Bockris, Appl. Phys. Lett. **55**, 4386 (1989)
12. C. Kainthla, B. Zelenay, and J. O'M. Bockris. J. Electrochem. Soc, , **133**, 248 (1986)
13. H. Morisaki, T. Watanabe, M. Iwase, and K. Yazawa, Applied Physics Letters., **29**, 338 (1976)
14. N. R. Tacconi, C. R. Chenthamarakshan, K. Rajeshwar, T. Pauporte, D. Lincot, Electrochem. Comm., **5**, 220 (2003)
15. S. Karuppuchamy, J-M. Jeong, D. P. Amalnerkar, H. Minoura, Vacuum., **80**, 494 (2006)
16. C. Natarajan, and G. Nogami, J. Electrochem. Soc., **143**, 1547 (1996)
17. A. Kay, I. Cesar, and M. Grtzel., J. Am. Chem. Soc., **128**, 15714 (2006)
18. N. Negishi, K. Takeuchi, and T. Ibusuki, J. Sol-Gel Scien. and Technol., **13**, 691 (1998).
19. S. U. M. Khan, J. Akikusa ,Electrochem. Soc. **145**, 89 (1998)

Low Reflectance Surface Observed on InP Porous Structures after Photoelectrochemical Etching

T. Sato, N. Yoshizawa, H. Okazaki, and T. Hashizume

Research Center for Integrate Quantum Electronics, Hokkaido University, North 13, West 8, Sapporo 060-8628, Japan

Extremely low reflectance was obtained from InP porous nanostructures in UV, visible, and near-infrared light ranges. The reflectance strongly depended on the surface morphology of the porous structures prepared by the electrochemical process, and the lowest reflectance of 0.1% in the visible light range was obtained from a sample after the irregular top layer was completely removed. Large anodic photocurrents were obtained on the InP porous structures that had low reflectance surfaces with deeper pores.

Introduction

InP is one of the most attractive materials for high-speed electronics and optoelectronic devices. Photosensitive devices such as photodetectors (PDs) in long-wavelength optical fiber communication systems (1,2) and high-efficiency solar cells with multi-junction structures (3) are excellent examples of technologies that utilize the superb optical properties of InP-based systems. With the aforementioned devices, however, surface reflection is a serious problem that degrades the device performance because is obviously reduces the efficiency of a photon energy conversion (4). Surface texturing of V-grooves (5)-(7), formation of insulator films (8), wide bandgap semiconductor films (9), and transparent conducting oxide (TCO) films (10) have been investigated as anti-reflective (AR) layers on top of InP.

In this paper, we report that extremely low reflectance below 0.4% was observed from the InP porous nanostructures in UV, visible, and near-infrared ranges. The surface reflectance spectroscopy was carried out on the porous samples prepared using the electrochemical process, leading to finding that the reflectance strongly depended on the surface morphology. Photoelectrochemical measurements on the porous structures revealed that the photocurrents increased in the samples that had low reflectance surfaces with deeper pores.

Experimental

The porous samples were electrochemically formed on a layer of (001) n-type InP (n=2-3×10^{18} cm^{-3}), on the back of which a GeAu/Ni-ohmic contact was formed for a current supply. The InP substrate was first anodized at 5 V in an HCl-based electrolyte to form the porous structures, where the structure depth, d, could be controlled by the anodization time, t_a (11). A disordered irregular layer formed and remained on top of the ordered porous layer after the first anodization. To remove this irregular layer, the porous surface was then photo-electrochemically etched under illumination at an anodic bias of 1 V in the same electrolyte (12). Figure 1(a) shows typical scanning electron microscope (SEM) images of the InP porous sample after photo-electrochemical (PEC) etching for 200 s (t_{PEC} = 200 s). The average pore diameter in the sample before the PEC etching was about 30 nm. On the other hand, regular sized pores about 130 nm in diameter appeared after PEC etching with t_{PEC} = 200 s, which was useful in completely removing the disordered layer from the surface. From the SEM observation on the cross-section, it was found that 130-nm-diameter nanopores were laterally separated by 50-nm-thick InP nanowalls.

In this study, the reflectance spectra of samples were first characterized using an ultra-violet (UV)-visible spectrometer (UV-1700, Shimadzu). The photon energy range of the light source was set from 1.12 to 6.0 eV corresponding to the wavelength range from 205 to 1100 nm. After that, the photoelectrochemical measurements were conducted on the various porous InP electrodes under white light using a tungsten lamp. The experimental setup and the schematics for the porous electrode are outlined in Fig. 1(b). A series of electrochemical measurements was conducted using a standard cell with three electrodes: an InP porous electrode used as a working electrode (WE), a Pt counter electrode (CE), and a saturated calomel electrode (SCE) used as a reference. All electrodes were dipped into 1mM $K_3Fe(CN)_6$ electrolyte. The potential of WE, V_s, with respect to SCE was precisely controlled with a potentiostat with a voltage source. In this electrolyte system, n-type InP surface is relatively stable and a large potential barrier with the $[Fe(CN)_6]^{3-}/^{4-}$ redox potential is expectable (13).

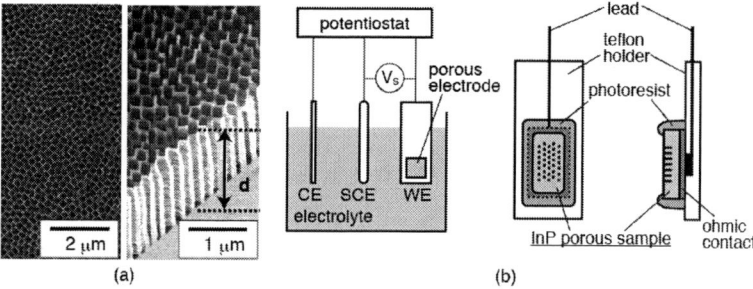

Figure 1. (a) SEM images of the porous sample after the complete removal of irregular top layers by PEC etching. (b) Schematic illustration of the electrochemical process and InP porous electrode used for the electrochemical measurements.

Results and discussion

We found that the optical view of the porous surface was reflected in the differences in the microscopic surface morphology. As shown in Fig. 2(a), the PEC etched porous sample has diffuse optical reflection resulting in a black surface. In order to clarify the optical reflectance on the porous structures, we investigated the effects of the irregular top layer on the surface reflectance by comparing three kinds of samples, as shown in Fig. 2(b). The pore depth, d, of the sample just after anodization without PEC etching was 13.8 μm, including the 4.0-μm-thick irregular top layer. After the sample was PEC etched for 200 s, the d decreased to 9.6 μm due to the irregular layer being removed. First, the reflectance of the planar sample was higher than 30% over the measurement range. Typical peaks were observed around 1.4 eV, 3.0 eV, and 4.5–5.6 eV, which were attributed to the interband transitions (14). The highest peak observed below 1.4 eV was the reflection on the back ohmic contact, which acted as a mirror for light transmitted through the InP bandgap. The reflectance obtained from the porous sample without PEC etching was 5–10% lower than that obtained from the reference sample. As plotted in Fig. 2(b), the spectral features of the InP bulk remained in the porous sample, showing that crystal quality was maintained beyond a certain level after the pore formed. On the other hand, the reflectance of the porous sample with the PEC etching drastically decreased compared with the other two samples. These results are very consistent with the optical view shown in Fig. 2(a). For the sample after the PEC etching, the reflectance was lower than 0.4% over the measurement range.

A possible explanation for the low reflectance observed here is that the straight nanopores with enlarged openings appeared on the surface after the PEC etching. In the present case with our porous samples, air holes are closely aligned between the InP nanowalls with a filling factor, f, of about 0.3. This kind of air-dielectric composite, which when assembled in an ordered array, has a small n close to unity leading to the low reflectance on the air interface. On the other hand, it should be considered that the irregular top layer has a large n value due to the winding pores with poor porosity, resulting in the surface reflectance reducing by only a few percent.

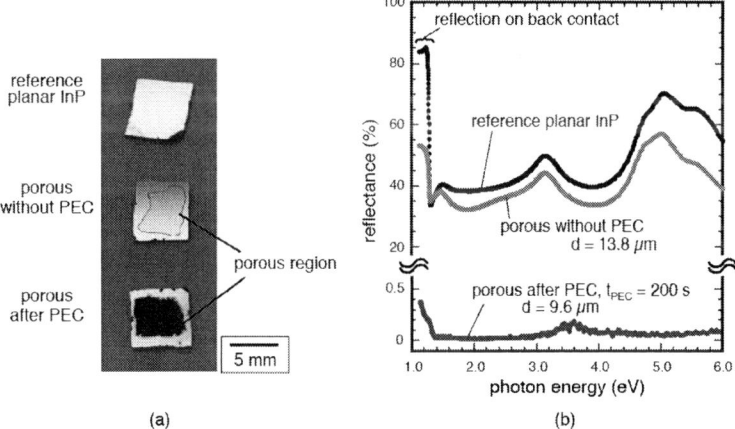

(a)   (b)

Figure 2. (a) Photograph of planar InP reference, porous sample without PEC etching, and porous sample after PEC etching. (b) Surface reflectance spectra measured as a function of photon energy of incident light for three samples.

The photoelectrochemical measurements were conducted on the InP porous structures after the irregular top layer had been removed. Figure 3(a) shows the cyclic voltammograms obtained at the porous InP electrodes with d=9.6μm in the $K_3Fe(CN)_6$ electrolyte. The current density was calculated using the geometrical surface area of the electrode. As shown in Fig. 3(a), the cathodic currents changed little with light illumination. However, the anodic currents increased under illumination where the dark current showed good blocking behavior. These results indicate that the basic photovoltaic property of n-type semiconductors remained on the porous InP electrode.

The anodic photocurrents of the various porous electrodes with different d's of 3.2, 4.8, 7.2, and 9.6 μm are compared in Fig. 3(b). In comparison, the photocurrents observed for the planar electrode (d=0 μm) were very small and negligible in a range of 0.1mA/cm². On the other hand, the photocurrents of the InP porous electrodes were observed in this current range for all samples, as shown in Fig. 3(b). Furthermore, the anodic currents were found to increase with d whereas the cathodic currents did not change very much.

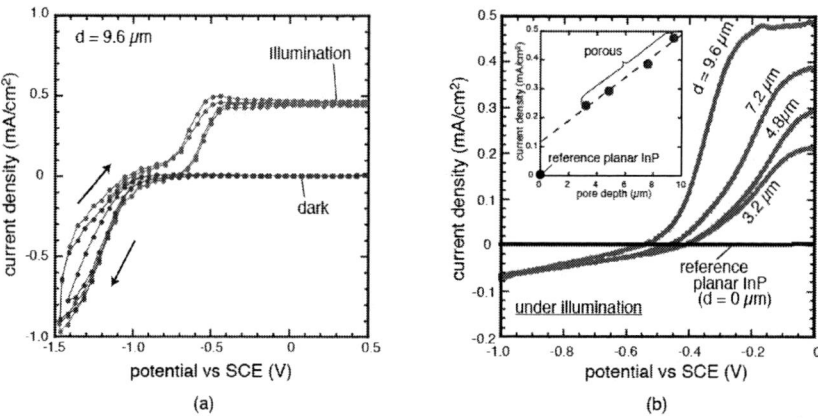

Figure 3. (a) Cyclic voltammograms measured on InP porous electrode with pore depth, d, of 9.6μm in $K_3Fe(CN)_6$ electrolyte. (b) Current-voltage (I-V) characteristics of InP porous structures that have different pore depth, d, of 3.2, 4.8, 7.2 9.6μm. Inset shows the correlation between anodic photocurrents measured at applied potential of 0V and d.

Inset of Fig. 3(b) shows the plot of the anodic photocurrents measured at the applied potential of 0V as a function of the pore depth, d. This plot showed approximately linear relationships with d as expected because the anode efficiency proportionally increased with the surface area of the inside pore that was enlarged as a linear function of d. Similar effects of the enlarged surface area have been reported on the amperometric chemical sensor based on the InP porous structures (15). However, the difference in the photocurrents between the porous electrodes and the reference planar electrode cannot be explained only by the effects of the enlarged surface area. As shown in the inset, the extrapolation value given by the data of the porous electrodes was larger than the experimentally obtained value plotted at d=0 μm for the reference planar electrode.

These results suggest that the porous structures have a larger incident light absorption property than bulk InP. As shown in Fig. 2(b), the extremely low reflectance surfaces were obtained on the porous structures after the PEC etching, where the pores with the enlarged openings were vertically aligned. In such a case, the incident light goes through the porous layer by repeating the absorption on the pore walls due to an effective optical path longer than that of bulk InP. Similar discussion has been also made on the light absorption process in nanowire materials (16). Another possible reason for the absorption enhancing is the optical transition involving both surface states (17) and quantum states (18) formed on the pore wall. These additional transitions give extra absorption in a wider energy range than the band gap of bulk InP. Thus, it can be concluded that the efficiency of a photon energy conversion is enhanced in the porous layer, resulting in the large anodic photocurrents increasing with the number of positive holes.

Conclusions

In conclusion, we have demonstrated that the surface reflectance of InP was significantly reduced by the formation of the porous structures. PEC etching after the pore formation very effectively reduced the reflectance to less than 0.1% in the visible light range and less than 0.4% in the UV and near-infrared ranges. Photoelectrochemical measurements on the porous structures revealed that the large anodic photocurrents are obtained on the samples that have low reflectance surfaces with deeper pores. These results can be explained by the two unique features of the InP porous structures: the large surface area inside pores and the large photon absorption in the porous layer.

Acknowledgments

The work reported here was supported in part by a Grant-in-Aid for young scientists (A) - 21686028 and by Grant-in-Aid for Challenging Exploratory Research - 21656078, from the Japanese Ministry of Education, Culture, Sports, Science, and Technology.

References

1. C. X. Shi, D. Grutzmacher, M. Stollenwerk, Q. K. Wang, and K. Heime: *IEEE Trans. Electron Devices*, **39**, 1028 (1992).
2. F. Zappa, A. Lacaita, S. Cova and, P. Webb: *Opt. Lett.*, **19**, 846 (1996).
3. A. Khan, A. Freundlich, J. Gou, A. Gapud, M. Imazumi, and M. Yamaguchi: *Appl. Phys. Lett.*, **90**, 233111 (2007).
4. M. F. Schubert, F. W. Mont, S. Chhajed, D. J. Poxson, J. K. Kim, and E. F. Schubert: *Opt. Exp.*, **16**, 5291 (2008).
5. P. Jenkins, G. A. Landis, N. S. Fatemi, D. Scheiman, X. Li, and S. G. Bailey: *Sol. Energy Mater. Sol. Cells*, **33**, 125 (1994).
6. N. G. Ferreira, D. Soltz, F. Decker, and L. Cescato: *J. Electrochem. Soc.*, **142**, 1348 (1995).
7. N. L. Dmitruk, O. Y. Borkovskaya, I. B. Mamontova, O. I. Mayeva, and O. B. Yastrubchak: *Thin Solid Films*, **364**, 280 (2000).
8. M. G. Boudreau, S. G. Wallace, G. Balcaitis, S. Murugkar, H. K. Haugen, and P. Mascher: *Appl. Optics*, **39**, 1053 (2000).
9. M. R. Hantehzadeh, M. Ghoranneviss, A. H. Sari, F. Sahlani, A. Shokuhi, and M. Shariati: *Thin Solid Films*, **515**, 547 (2006).
10. K. Ramamoorthy, K. Kumar, R. Chandramohan, K. Sankaranarayanan, R. Saravanan, I. V. Kityk, and P. Ramasamy: *Opt. Commun.*, **262**, 91 (2006).
11. T. Sato, T. Fujino, and H. Hasegawa: *Appl. Surf. Sci.*, **252**, 5457 (2006).

12. T. Sato, and A. Mizohata: *Electrochem. Solid State Lett.*, **11**, H111 (2008).
13. A. Theuwis, and I. E. Vermeir: *J. Electrochem. Soc.*, **146**, 1172 (1999).
14. P. Lautenschlager, M. Garriga, and M. Cardona: *Phys. Rev. B*, **36**, 4813 (1987).
15. T. Sato, A. Mizohata, N. Yoshizawa, and T. Hashizume: *Appl. Phys. Exp.*, **1**, 051202 (2008).
16. O. L. Muskens, J. G. Rivas, R. E. Algra, E. P. A. M. Bakkers, and A. Lagendijk: *Nano Lett.*, **8**, 2638 (2008).
17. H. Fujikura, A. Liu, A. Hamamatsu, T. Sato, and H. Hasegawa: *Jpn. J. Appl. Phys.*, **39**, 4616 (2000).
18. T. Sato, T. Fujino, and T. Hashizume: *Electrochem. Solid State Lett.*, **10**, H153 (2007).

Author Index

Adachi, M.	1	Turner, J. A.	49
Bellardita, M.	29	Wang, H.	49
Bouttemy, M.	21		
		Yoshizawa, N.	83
Di Paola, A.	29		
Enomoto, M.	37		
Etcheberry, A.	21		
Feuillet, G.	21		
Frites, M.	73		
Hashizume, T.	83		
Ihara, M.	37		
Ingler Jr., W.	73		
Joudrier, A.	21		
Kawakita, J.	9		
Khan, S. U.	73		
Levy, F.	21, 21		
Mori, Y.	1		
Neophytides, S. G.	63		
Okazaki, H.	83		
Palmisano, L.	29		
Parrino, F.	29		
Sato, T.	83		
Seferlis, A. K.	63		
Simon, N.	21		
Taniguchi, K.	37		